National Systems of Innovation

National Systems of Innovation

Creating High-Technology Industries

Stuart Peters

First published 2006 by
PALGRAVE MACMILLAN
Houndmills, Basingstoke, Hampshire RG21 6XS and
175 Fifth Avenue, New York, N. Y. 10010
Companies and representatives throughout the world

PALGRAVE MACMILLAN is the global academic imprint of the Palgrave Macmillan division of St. Martin's Press, LLC and of Palgrave Macmillan Ltd. Macmillan® is a registered trademark in the United States, United Kingdom and other countries. Palgrave is a registered trademark in the European Union and other countries.

ISBN-13: 978–1–4039–4614–0   hardback
ISBN-10: 1–4039–4614–0   hardback

This book is printed on paper suitable for recycling and made from fully managed and sustained forest sources.

A catalogue record for this book is available from the British Library.

Library of Congress Cataloging-in-Publication Data
Peters, Stuart, 1965–
    National systems of innovation : creating high technology industries / by Stuart Peters.
        p. cm.
    Includes bibliographical references and index.
    ISBN 1–4039–4614–0 (cloth)
        1. Semiconductor industry–Government policy–United States.
    2. Semiconductor industry–Government policy–East Asia. 3. Semiconductor industry–Government policy–Europe. 4. Liquid crystal display industry–United States. 5. Liquid crystal display industry–East Asia. 6. Liquid crystal display industry–Europe. I Title.

HD9696.S43P48 2005
338.4′76213815–dc22   2005051454

10   9   8   7   6   5   4   3   2   1
15   14   13   12   11   10   09   08   07   06

Printed and bound in Great Britain by
Antony Rowe Ltd, Chippenham and Eastbourne

To all my friends and family

Contents

List of Figures

Acknowledgements

The writing and completion of this book has been a long and hard road. Its origins date back to when I was doing my research for my PhD whilst at Brunel University. I was initially put on the intrepid road following suggestions from my two former supervisors, Professor Martin Cave (now at Warwick University) and Dr John Howells (now at Aarhus University) that I had something worth publishing in my possession. I received additional encouragement from Dr Anne-Marie Coles and Professor Keith Dickson (Brunel University), Dr Vinita Damodaran (University of Sussex) and Dr Richard Grove (Australia National University) to 'bite the bullet' and write the book. There were several times I was unsure if it was worth all the effort. Now the book has been completed I am glad to say that I am more than grateful for their initial encouragement and advice. Lastly, I would like to extend my thanks to Mr Lakis Kaounides (Queen Mary, University of London) for some very timely and useful suggestions on the final product. The usual disclaimer applies.

Introduction

It is widely known today that the rise of Japan as an economic super-power, particularly during the 1980s, had massive reverberations. These reverberations were felt most acutely in the US and Western Europe as their once dominant positions across a range of industrial sectors came under a sustained attack of a type that they had never witnessed before. Sectors such as steel, automobiles, and electronics, in other words, both mature and high-technology industries, were attacked simultaneously. Japan's success precipitated an enormous amount of 'soul-searching' into what Japan was doing 'right' and the US and Western Europe were doing 'wrong'. This soul-searching reached fever pitch in the US with the publication of several government and private sector reports, such as the Department of Commerce *Emerging Technologies: A Survey of Technical and Economic Opportunities* (1990), the Department of Defense *Critical Technologies Plan* (1990), and the (privately funded) US Council on Competitiveness *Gaining New Ground: Technology Priorities for America's Future* (1991). This was in addition to the influential book *Made in America* by Dertouzos *et al.* (1989). After enjoying more than three decades of almost unrivalled economic dominance, the foundations on which it had been built developed serious fault lines. The challenge to its economic dominance, and more specifically, its technological supremacy was very unwelcome. The country had come to believe its technological supremacy was a birthright. To the obvious frustration of the US, it made the assumption that the 'rules of the game' had been rewritten under its nose and the only way Japan could have caught up was by some underhand method.

The method identified was industrial targeting. From the perspective of the US, the very success of this type of policy made it 'unfair' as the targeted sectors were often the very same ones where the US had significant competitive strengths. Japan is not the only country to have actively promoted key sectors. Western Europe, and other East Asian countries, notably South Korea have also followed suit. The area of greatest sensitivity was in high-technology industries which the US considered a domain all of its own. Not surprisingly, a debate erupted in the US over how it should meet this new challenge. The ensuing debate was most illuminating as it focused upon the extent to which the US could 'intervene' to preserve and solidify its dominant position.

The debate was a polarised one between advocates who contended that the emergence of competitors was inevitable, and those who argued that intervention was necessary to arrest the situation. Nelson (1990) held the former viewpoint. As his analysis reveals, in a paper with the very apt title, 'US Technological Leadership: Where Did it Come From and Where Did it Go?', there were several factors at work over which the US had no control enabling her to constrain the emergence of competitors. Two of these are especially noteworthy. Firstly, the globalisation of technology made it much easier for countries to obtain the technology they wanted; and secondly, an increasing number of countries had the capacity to make substantial investments to build up a presence in high-technology industries. From Nelson's standpoint, the only way the US could fend off its new competitors was to take a number of domestic policy measures to restore its own competitiveness. In light of the prevailing notions against any form of 'industrial policy' at the time, intervention was not a viable proposition. Efforts to emulate some aspects of Japanese strategies in creating joint R&D programmes, for example, the Flat Panel Display Initiative, came later under the Clinton administration.

On the other side of the spectrum were Krugman (1986) and Tyson (1992) who argued intervention was warranted on the grounds that the loss of technological leadership should be reversed. Government intervention in industry is normally equated with, for example, channelling subsides into a declining industry. In the 1980s intervention was given a new dimension when the US decided it would not follow conventional wisdom and intervene in the industrial systems of individual countries. As explained later in Chapter 1, this became known as 'cautious activism'. Basically, the logic of the strategy was to dent the competitiveness of its competitors in such a way that the US could regain its former momentum. The characteristic features of high-technology industries; imperfect competition, strategic behaviour, dynamic economies of scale, and technological externalities (Tyson, 1992:4), provided the US with the justification for this new form of intervention. Such was the level of concern about the loss of technological leadership across a range of technologies that the advocates of cautious activism won the day. The US put itself on a course which it believed would 'regain' its technological leadership.

Throughout the debate on the virtues of interventionism, one crucial element was missed out in its entirety. No proper consideration was given to *how* high-technology industries are created. The fundamental weakness of industrial targeting is it gives a highly misleading impression; all a country has to do if it wants to create a high-technology industry is to wave a 'magic' policy wand. Analyses such as those of Okimoto *et al.* (1984) and Anchordoguy (1989) were sidelined. These analyses highlight the difficulties involved in the development of high-technology industries. At the same time they also suffer from a major shortcoming. They do not help explain why certain countries have proved more adept at developing high-technology industries than others. To resolve this conundrum, it is necessary to turn to Freeman's (1987) concept of the national system of innovation (NSI). Ultimately, it is a country's NSI that determines whether it can develop the high-technology industries of its own choosing, and not the 'simple' waving of a policy wand that governments have a preponderance to do.

Through a synthesis of a wide range of literature and empirical evidence drawn from the fields of management, economics and innovation, the book illustrates the complexity of developing two specific high-technology industries, semiconductors and liquid crystal displays, a characteristic that is rarely recognised. Furthermore, the focus is a global one. In an era of globalisation, it makes no sense to only analyse the different strategies of the US and Japan in high-technology industries. It is now necessary to incorporate Western Europe, South Korea, and Taiwan as they also often compete in the same high-technology industries as the US and Japan. Any attempt to understand the changing dynamics of high-technology industries without their inclusion is otherwise impossible. The inclusion of Taiwan and South Korea is especially important, as they have in recent years put Japan under major competitive pressures in addition to the US. Compared to the 1980s and early 1990s, much has changed to the extent that Japan is no longer the same fearsome competitor in high-technology industries it once was. However, new complex variables have now entered the dynamics of these developments. These include the rise of China (and its targeting of high-technology industries) and India, as well as the emergence of major generic new technologies such as nanotechnology, biotechnology, and information technologies, and, most importantly, their convergence.

1
'Merging Together': The Semiconductor and Liquid Crystal Display Industries

The explosion of the 'information age' in the 1990s has transformed many consumer markets and areas of the business environment way beyond anything which would have been considered remotely possible in the previous decade. It requires no further elaboration here to emphasise how far-reaching this technological revolution has been. From its very humble origins in the 1970s with simple computer generated credit-card statements and household bills, the diffusion of automated teller machines (ATMs) (Davis & McCormack, 1979),[1] to the widespread adoption of the facsimile (FAX) machine in the 1980s, the information age has undergone a period of phenomenal growth and change. It has been driven by four main factors; the emergence of the World Wide Web, the spread of high-speed Internet access, the rapid progress in personal computing technology, and the development of the digital infrastructure.[2] These major developments have had a profound effect in a number of different ways. Huge amounts of data blending text, sound and images can now be transmitted at lightning speed and efficiency and once distinct industry boundaries have become blurred.

An excellent illustration of the seismic transformation that has occurred can be seen in the 'infoimaging' industry, a new hybrid industry created through the convergence of image science and information technology.[3] This industry has grown out of the increasing use of colour in documents. Whether the documents are produced individually, downloaded straight from the Internet, or for that matter photographs routed to printers and copiers, it has required the construction of a new infrastructure (photofinishing systems and networks, processing equipment) and created a huge demand for a new range of products (inkjet printers, flat-bed scanners, digital cameras, flat panel displays (FPDs)) and services (photographic processing, film and paper, document and health imaging, inkjet papers and cartridges). In 2001 one estimate put the global value of this hybrid industry at approximately $225 billion, with photographic film, papers and processing worth $54 billion alone.[4]

Little is it realised that the information age would not have materialised without the tremendous technological advances made in related industries. Two of the industries at the inception of the information age have been semiconductors and liquid crystal displays (LCDs). The enormous dynamism these industries have displayed has raised fundamental questions about what are the factors which support and power these two high-technology industries at an unprecedented rate and have given them such a high degree of complexity.

The semiconductor industry has been in existence for more than five decades and has a long established reputation for making important technological breakthroughs, mainly pioneered in the US, such as the development of the integrated circuit (IC) and microprocessor. The industry has continued to go from strength to strength as more and more entrants have entered the industry, forcing those already in it to continually push forward the technological frontiers to survive and remain competitive. The changing dynamics of the semiconductor industry has now fed through into a closely related industry, LCDs. The LCD industry has only emerged as a major industry within the last decade. The two industries are linked together in a number of subtle ways. Firstly, semiconductors and LCDs can be found in many of the same products, and secondly, they are both very complex to develop and manufacture. Lastly, the countries which have been responsible for the recent breathtaking pace of developments in the semiconductor industry are exactly the same ones which have also been at the forefront of developing the LCD industry. This is all the more extraordinary given that it is the countries of East Asia which now dominate the LCD industry even though the early technological lead was initially held by the US. The contribution of these countries to the development of the LCD industry cannot be overestimated. In growth terms, they have driven the industry forward at such a phenomenal rate that it has even been suggested by Murtha *et al.* (2001) that the pace of the development of the LCD industry has been more than double or even quintuple the development pace of semiconductors during the analogous period in that industry's history.[5] A virtuous circle of innovation has been created whereby the industries have reached the stage where they now fuel each other, in other words, advances in one industry stimulate further advances in the other. For the countries which are simultaneously involved in these two industries it has brought them rich rewards and looks all set to continue to do so well into the future.

In most major industrial sectors, it is true to say that there is normally a substantial contingent of European and US firms present, whether it is in steel, machine tools, or electronics. However, what is most unusual about the LCD industry is that the numbers of European and US firms present, when compared to other sectors, is absolutely minuscule. This could be understandable if the LCD industry was a 'mature' industry, where European and US firms had been forced to make an exit because of 'low-cost' competition, especially from firms in the Far East. The fact of the matter is, however, that the LCD industry is a high growth, capital intensive, high-technology industry. Apart from semiconductors, there are virtually no other products that are so technologically advanced. Furthermore, in the last decade LCDs have become increasingly important in their own right because of their rapid proliferation. Against this background, it is therefore very surprising to find that European and US firms have not become major players in this rapidly growing and technologically advanced industry. This immediately leads on to the question, why?

It should come as no surprise that to answer this question one has to look no further than Japan, which was the only other country in a similar position to Western Europe and the US during the 1970s to exploit the 'potential' of LCDs, to see where they have gone wrong. Ever since the early 1970s Japan has been at the forefront of developing LCD technology, unlike Western Europe and the US, which did not view LCD technology as being important until the late 1980s. Suffice to say here, by the early 1990s Japan's decision to develop LCD technology had begun to yield real dividends. It had a monopoly in thin-film transistor (TFT) LCDs, which are the third generation of LCDs.[6] Japan's global market share was estimated to have been approximately 90–100 per cent. In the following decade, competition in the industry has become increasingly intense. The once seemingly impenetrable Japanese stranglehold has been successfully broken by South Korea and Taiwan, though the efforts made by the US and Western Europe to make substantial headway in the industry has come to little. The success of Japan, South Korea, and Taiwan in the LCD industry suggests they have established a major competitive advantage over the US and Western Europe, which is a most peculiar situation. Furthermore, it directly challenges the highly questionable contention made by Murtha *et al.* (2001):

> The FPD industry [LCD industry] represents an example of a new class of industry – a class that includes software, e-commerce, and multimedia – that cannot be understood in terms of country rivalry or national competitive advantage. In these knowledge driven industries, learning rather than product positioning powers competition. The knowledge creation process that fuels these industries cannot be segregated by national territory.[7]

The argument put forward by Murtha *et al.* contains an important and inherent flaw; they 'concede' Japan was the dominant force in the LCD industry before the mid-1990s and suggested there was nothing 'special' about Japan:

> The FPD industry, like the high-tech industries in Silicon Valley, appears to have thrived on the mutual geographic proximity of its members. But it is important to emphasise that the industry's early concentration in Japan arose as a consequence of knowledge accumulation driving the technology forward and not the other way around. There was nothing intrinsic about Japan, Japanese management style, or any other Japanese business, academic, or government institution that uniquely destined Japan to serve as the global centre of the industry.[8]

Furthermore, Murtha *et al.* point out that South Korea and Taiwan have made a successful entry into the industry and have reached the point where they can compete with Japan on equal terms or better. This 'neutralises' their own assertion that the LCD industry cannot be understood in terms of country rivalry or national competitive advantage. The approach adopted by Murtha *et al.* to analyse the LCD industry has been to examine it from the perspective of the level of the firm. The justification used for this approach is that the development of the LCD industry has been contingent upon the *continuity, speed, and learning* of the firms involved. There are obvious merits to this approach, for example, it is important to understand how firms manage the rapid pace of technological change that is characteristic of the industry. The approach, however, has a major deficiency which is critical to the development of a more holistic assessment of the LCD industry; it takes no account of the *technological capabilities* of a country or group of countries. This is necessary to begin to explain the wide variations in the performances of Japan, South Korea, Taiwan, the US and Western Europe. The rapid emergence and growth of the LCD industry has brought into sharp focus the importance of why a country should maintain its technological capabilities. With the 'appropriate' capabilities it can successfully develop this *type* of industry, without them, it can end up with nothing.

Technological dreams and realities

For more than three decades, it has been the dream of engineers worldwide to develop a 'wall hanging' television, that is, a television that can literally be hung on a wall like a painting. Engineers have known for a very long time that to have any chance of turning this 'concept' into reality a viable alternative had to be found to the cathode-ray tube (CRT). It is the CRT that gives the conventional television set the capacity to produce high quality pictures. The CRT has a number of important advantages, which is why it has been the dominant display technology for as long as it has. That said, it should be borne in mind that the CRT also suffers from serious deficiencies which makes it vulnerable to challenges from other display technologies, of which there are several.

The CRT's main advantages are that it is cheap to manufacture, offers good brightness, contrast, colours, resolution, and reliability. The major problem with the CRT is that it is power-hungry, as well as being extremely heavy and bulky. The main applications for the CRT, and where it has enjoyed unrivalled dominance for several decades, have been in televisions, and more recently, in computer monitors. For applications that need displays that are thin, extremely lightweight, and consume very little power, the CRT is of no use. This is where other display technologies, called flat panel displays or FPDs come into their own.

Until the mid to late 1990s FPDs remained in relative obscurity. Up to that point the various different types of FPDs were largely unknown because they were still undergoing development and were nowhere near ready for the marketplace. There are a number of different FPD technologies: plasma, vacuum fluorescent, light-emitting diodes (LEDs), light-emitting polymers (LEPs), electroluminescence, field-emission displays (FEDs), and lastly, LCDs. Some of these technologies have been around for more than three decades, such as the LED, LCD, electroluminescence and plasma, but it is the LCD that has emerged as the dominant FPD technology. The LCD has emerged as the dominant FPD technology because as more development work was undertaken on the various technologies in the 'race' to develop a wall hanging television, so more became known about the strengths and weaknesses of the respective technologies. In the 1960s, for example, both electroluminescent and plasma technologies were held up as promising technologies, but as it has turned out the LCD has eclipsed them both, and has proved the most versatile by far. The LCD might have emerged as the dominant FPD technology but it does not mean to say that either electroluminescent or plasma technologies have disappeared from the scene, far from it. As a direct consequence of the race to develop a wall hanging television, the early promise that plasma technology showed in the 1960s has finally come to fruition. It is now one of the dominant technologies in the use of wall hanging televisions currently being produced by firms mainly from East Asia.

Going back to LCDs, since Sharp launched a calculator incorporating an LCD onto the market in 1973, LCDs have become widespread and commonplace. Today, LCDs can be found in a host of products ranging from the traditional watches and calculators to road traffic guidance systems, photocopiers, mobile phones, camcorders, notebooks and computer monitors. In other words, the LCD has helped bring about major technological improvements in established products, whilst at the same time helped spawn an entirely new generation of consumer and industrial products. The great majority of people take the products listed above, and others like them, for granted as it is all too easy to forget just how quickly they have emerged and become a vital part of our everyday life. Behind this rapid proliferation of LCDs, however, is an industry that has quickly grown into one of global importance. But to appreciate its peculiar nature, and analyse what lessons can be learnt from its development, it is necessary that it is closely compared with the semiconductor industry.

Semiconductors and LCDs

In semiconductors, the US was the first country to build up a significant technological lead in the industry. It took Japan approximately two decades to catch up with the US before it was in a position to successfully challenge it. Wong & Mathews (1998) assert that Japan's success in semiconductors, and other industries such as televisions, has given rise to a popular account of how Japanese firms allegedly build up a dominant position in an industry.[9] According to this 'stylised' account,[10] Japanese firms build up a dominant position in a 'selected' industry firstly by 'copying' a technology (one that is normally invented in the US and commercialised by US firms) with the active support of the Ministry of International Trade and Industry (MITI). MITI then helps to raise the technological capabilities of the competing Japanese firms through a combination of 'controlled' competition, a MITI-inspired R&D consortia, the sharing of resources, and the exchange of information. As the industry begins to expand, Japanese firms gradually increase the competitive pressures on US firms by raising the level of their investment and focusing on product and process innovations. The competitive pressures on US firms quickly snowball and soon reach a point where US firms are forced either into bankruptcy or have to make a 'strategic' exit, paving the way for Japanese firms to dominate the industry.[11]

As we shall see in the following chapters, what has occurred in the LCD industry over the past three decades bears only a partial resemblance to the account given above. Suffice to say here, that the US was the first country to develop a technological lead in LCDs, but unlike in semiconductors, it failed to build upon its early advantage. During the early stages of the industry, the US allowed Japan to take over the number one spot, a position it has maintained for three decades. What makes the LCD industry so unusual is how Japan took the lead. It did so, not by 'copying' a technology that was originally invented in the US and commercialised by US firms, but because of turmoil in the US digital watch industry in the late 1970s. This turmoil dealt a devastating blow to the US LCD industry, but Japan's LCD industry emerged from it virtually unscathed. In the 1980s, the US LCD industry continued to contract. Japan's LCD industry, on the other hand, forged ahead.

From the above, there are a number of issues that need to be addressed. The two most fundamental ones are: what led Western Europe and the US to 'dismiss' the importance of LCD technology until the late 1980s? And what are the specific factors that have enabled South Korea and Taiwan to catch up with Japan in TFT-LCDs, but not Western Europe and US? These issues are analysed in the following chapters, which have been structured to show how one 'outcome' is linked to the next, in one degree or another.

Chapter 2 focuses on national systems of innovation and path dependency as the framework to understand the variations in the performance of Japan, South Korea, Taiwan, the US and Western Europe in TFT-LCDs. Furthermore, it is argued that in an era of globalisation the national system of innovation (NSI) remains as important as ever. No country can expect to maintain and generate new technological capabilities without its NSI in 'good health'.

Chapter 3 analyses the different public policy frameworks of Japan, South Korea, Western Europe and the US. It shows that while they all have policies relating to competition, trade and technology, some policies are far more important to certain countries than others. For instance, Japan has always placed much more emphasis on its technology policy than either its trade or competition policies. At the other end of the spectrum, the US has always placed far more emphasis on its trade and competition policies than it has ever done on its technology policy. Not only that, there is also significant variation between the various countries in terms of how they pursue their respective policies. From Chapter 3, two other important points emerge. Firstly, that Western Europe and the US have modified their public policy frameworks in very specific areas so they are more in keeping with Japan's. They have done this as part of their wider attempts to improve their competitiveness vis-à-vis Japan. Whilst Western Europe and the US have been keen to 'copy' Japan in particular aspects, Japan and South Korea have interestingly shown no enthusiasm whatsoever to copy from Western Europe or the US. Secondly, despite the various changes that Western Europe and the US have made to their public policy frameworks, there has been no visible improvement in their competitiveness, at least not in the field of TFT-LCDs.

Chapter 4 analyses the developments in the semiconductor industry up to the early 2000s, and particularly focuses on the changes of leadership that the industry has witnessed over the last two decades. It begins by looking at the origins of the industry and how the 'initial conditions' gave the US an early important lead. It then goes on to examine the various factors that helped undermine the US position, and enabled Japan to catch up and then overtake the US. Since Japan overtook the US in semiconductors, especially in DRAMs (dynamic random access memories), the US has actively used its trade policy against Japan to try and re-establish its former presence in the industry. After more than a decade of trying, the US has made no progress on this front. The 'vacuum' created by the exit of a large number of US firms from the industry has since been filled by firms from South Korea. This confirms beyond any doubt that changes made to, or the use of, specific policies do not automatically lead to an improvement in a country's competitiveness in a key sector.

Chapter 5 analyses two mainstream but different strategies commonly pursued by firms, mergers and acquisitions and core competencies, to gain a competitive advantage. It assesses why some firms are more successful at learning about technology than others.

Chapter 6 lays the foundations for understanding about the 'peculiarities' of the LCD industry. It starts off by giving the historical background to how liquid crystals were 'discovered' and why it was that the US found itself, yet again, with an important early lead in what has evolved into a very important technology. This is followed by an analysis of the events which eventually led the US to 'give up' LCD technology, and presented Japan with a golden opportunity to exploit the technology, without having to face the additional worry of having to catch up with the US once more. The main conclusion drawn from the chapter is that at the end of the day what led the US to 'abandon' LCD technology was short-termism, which was then endemic amongst the firms in the industry.

Chapter 7 examines the strategy used by Japan to establish itself in a dominant position in the LCD industry. As mentioned earlier, Japan adopted a long-term view of LCD technology from the beginning of the 1970s. Japanese firms, most notably Sharp, set about establishing R&D facilities and long-term relationships with other firms interested in LCD technology. US firms, on the other hand, did nothing of the sort. These differences in strategies did not manifest themselves in any way during the 1970s. Even after the turmoil in the US digital watch industry, US firms maintained their short-term view of LCD technology. They did not even attempt to reverse the damage that had been inflicted on them by the turmoil in the US digital watch industry. Had US firms changed their view towards LCD technology and followed the example set by Sharp, then competition for the leadership of the industry would probably have been very different. By the early 1980s, Japan had made substantial advances in LCD technology. It was the undisputed leader of the industry, a position that it enjoyed until the mid to late 1990s when the first serious challenge to its leadership emerged.

Chapter 8 examines the principal reasons why South Korea has caught up with Japan in TFT-LCDs. The underlying explanation is straightforward, but at the same time is quite complex in its own way. It is argued that because South Korea has deliberately tried to emulate the Japanese 'model' of economic development, this has given the country several important features that Japan also possesses. These features have proved highly advantageous when it comes to developing capital intensive, technologically advanced industries, most notably DRAMs and TFT-LCDs. Like Japan, South Korea has a strong presence in DRAMs, and part of its success in TFT-LCDs can be attributed to this, as the two industries have close links. Apart from South Korea, the only other country that has become a major player in the TFT-LCD industry is Taiwan. Unlike South Korea, which rushed into the industry with little regard for the high risks that are inherent in this type of industry, Taiwan has adopted a much more cautious approach. It has chosen to look to Japan in a big way to help it develop a presence in the TFT-LCD industry. Japan has shown a considerable degree of willingness to assist Taiwan in recent years, and this has provided Taiwan with the expertise to build up its own technological capabilities.

Chapter 9 analyses the 'failed' strategies of the US and Western Europe to establish a presence in TFT-LCDs. Although the two strategies are very distinct from each other, they both failed to take into account one crucial element, the importance of production, which helps to explain why the US and Western Europe have found it so hard to make a successful entry into the industry. Production is, of course, an area where Japan has long excelled. Until Western Europe and the US start to appreciate the importance of production, and take actual measures to improve their performance in this area, they can continue to expect to struggle in industries such as TFT-LCDs.

Chapter 10 draws together the themes of the book and questions raised.

With the rapid emergence of the TFT-LCD industry as a major industry in its own right during the 1990s, further investigation and analysis is warranted into the one industry that has really tested the industrial muscle of Western Europe and the US to the full, and found it severely wanting. Although Western Europe and the US have experienced major competitive problems in the semiconductor industry over the past two decades, they have managed to come through many of these difficulties and remain major players within the industry. When the US started to experience major problems in semiconductors during the 1980s, the US alleged that Japan was guilty of using 'unfair' trading practices to gain a competitive advantage in the industry. Japan's initial domination of the TFT-LCD industry had nothing to do with any so-called unfair trading practices, but was the combination of the country's technological capabilities and its long-term commitment to develop LCD technology.

The semiconductor industry provides a graphic illustration of the need to show why 'unfair' trading practices have nothing to do with Japan's success in TFT-LCDs. It is discussed later in Chapter 4 that since 1986 the US has tried to 'manage' the semiconductor industry with Japan through a series of agreements that it has forced upon the Japanese government and firms alike. The example set by the US has been followed to some extent by Western Europe. Whenever there is a downturn in the semiconductor market, Western Europe has shown a strong tendency to impose anti-dumping duties on semiconductors from Japan and South Korea to 'protect' European firms from 'unfair' competition.

The extensive intervention of the US government in the semi-conductor industry since the mid-1980s has been motivated by the very strong belief within the US that Japan had 'manipulated'[12] the industry for its own ends, using a variety of measures, at huge cost to US firms. The source of this increasing willingness of the US to inter-vene in high-technology industries such as semiconductors, comes from the idea of 'cautious activism'. Two of the leading proponents of this idea are Krugman (1986) and Tyson (1992).[13] They have pointed out that when it comes to competition between countries in high-technology industries there is no such thing as a 'level playing field'. The supporters of cautious activism argue that if the US is to maintain its position in high-technology industries, the country has to adopt a wide series of measures. Some of these measures are much more controversial than others. Measures such as the need to increase the public funding of education and civilian R&D, and a more generous R&D tax credit barely generate any controversy at all.[14] It is in the field of trade though where cautious activism generates a substantial amount of controversy.

Tyson (1992) points out that overall the US tries to have non-discriminatory trade policy, which amongst other things, seeks to main-tain open markets and open up markets that remain closed. In its trade relations with other countries, the US expects reciprocity, which is not an unreasonable expectation to harbour. Trouble starts to manifest itself when the US comes to the conclusion that the reciprocity it expects from a trading partner is not forthcoming to the extent that it finds 'acceptable'. Whenever a country falls foul on this point, the US will often make a number of demands to make the relationship between the two parties more equitable. Should the country not agree to these demands, or move at a pace that is too slow for the liking of the US, the US then feels with some justification that it has the right to 'act'.

When it acts the US usually takes one of two courses of action. It either threatens to deny access to the US market, or it makes some sort of bilateral 'agreement' which goes a long way to meeting its original demands. When Tyson states that 'cautious activism does sometimes involve forceful unilateralism',[15] it really does mean just that.

It is this willingness to engage in unilateral action that has led to the situation which now exists in the semiconductor industry. Cautious activism was used to deal with a specific problem in the semiconductor industry in the mid-1980s, but even after the original problem has long since 'dissipated', cautious activism is still very much in evidence in the industry. The experience of the semiconductor industry suggests that once it has taken root, cautious activism is very difficult to eradicate. The semiconductor industry shows that at certain moments in time government intervention is warranted, however, intervention is only useful up to a point. Whether any government likes it or not, the fact of the matter is that what determines whether a country remains competitive in an industry is how innovative and nimble its firms are.

Admittedly, the semiconductor industry is very different from the TFT-LCD industry in the sense that the US has an industry to 'protect', but in TFT-LCDs it has no industry of a comparable size and importance to protect. The point is that any use of cautious activism of the type witnessed in the semiconductor industry will not create a domestic TFT-LCD industry which the US 'requires', nor will it fundamentally change the technological position of Japan and South Korea. Even Florida & Browdy (1991)[16] and Tyson (1992) have to concede that the failure of the US to have a TFT-LCD industry lies with the US itself, and that trade policy alone cannot solve the situation. The country's failure to tackle its 'production problem' is its own responsibility and no one else's. Whether policymakers in the US will ever come to appreciate how extensive the production problem is, and seek to remedy the country's lack of a TFT-LCD industry through the use of trade policy is anyone's guess.

The book shows that for a country to thrive in the TFT-LCD industry requires the successful interaction of a variety of different factors. Should any one of these factors be lacking, then a country can expect to struggle in the industry as Western Europe and the US have shown. In theory catching up sounds relatively simple, but in practice it has a habit of proving fiendishly hard. Successful catch-up is highly dependent on a country's own indigenous technological capabilities. Should they be found to be defective, a country stands little hope. Catch-up appears 'easier' for countries that are in the process of experiencing rapid industrial development, but for countries that are already highly industrially developed the catch-up process has proved to be a major ordeal. Perhaps this should come as no surprise. When a country which has never been in the position of having to catch up is one day suddenly confronted with a situation where it has to, the chances of it being successful are very small because its industrial system has not been geared up to meet the challenge. This is the perspective which is adopted here.

In addition, the book aims to illuminate just how complex and different each of the countries' industrial systems are. Once a country has a particular type of industrial structure it seems to take on a 'life of its own'. What is meant by this is that any attempts to fundamentally reform the system, perhaps to correct any 'weaknesses' that may have begun to manifest themselves, are likely to prove more than an uphill struggle. Whether it is trying to change the behaviour of private firms, or the workings of financial institutions for the 'good' of the country, efforts are likely to be met with a substantial amount of resistance. Those firms (or financial institutions) which may happen to fall under the spotlight will always argue that they have to be free to do whatever is necessary to remain competitive in the sector(s) in which they operate. Convincing them to do otherwise is likely to fall on deaf ears. Policymakers might 'know' what is in the best interests of a country but actually making it happen is another story altogether.

Summary

This chapter has both described what the book is about, and discussed at some length the background to it. From the preceding discussion, two points particularly stand out. Firstly, that the growth of the LCD industry is a by-product of a wider technological race. It was the race to develop a wall hanging television that was to prove a major catalyst by helping to lay the foundations for the development of the present day industry. To recap, a number of different FPD technologies were developed simultaneously in an effort to turn this concept into reality. One of the main outcomes of the development process was that the LCD emerged as the most flexible FPD technology, which has subsequently grown into a major new industry of global importance. Secondly, close parallels exist between the semiconductor and LCD industries that cannot be ignored.

Notes

1. Davis, WS & McCormack, AM (1979) *The Information Age*, Reading, Massachusetts: Addison-Wesley.
2. Murtha, TP, Lenway, SA, Hart & JA (2001) *Managing New Industry Creation: Global Knowledge Formation and Entrepreneurship in High Technology*, Stanford, California: Stanford University Press, p. 4.
3. 'In Blur of Text and Image, Hybrid Industry Emerges', *International Herald Tribune*, 18 April 2001.
4. *Ibid.*
5. Murtha *et al.*, 2001:44.
6. Hung, SC (2002) 'The co-evolution of technologies and institutions: a comparison of Taiwanese hard disk drive and liquid crystal display industries', *R&D Management*, Vol. 32, No. 3, p. 185.
7. Murtha *et al.*, 2001:4.
8. *Ibid*:38.
9. Wong, PK & Mathews, JA (1998) 'Competing in the Global Flat Panel Display Industry: Introduction', *Industry and Innovation*, Vol. 5, No. 1, p. 2.
10. This is the term used by Wong & Mathews.
11. Wong & Mathews, 1998:2.
12. This is the term used by Tyson (1992).
13. Krugman, P (ed.) (1986) *Strategic Trade Theory and the New International Economics*, Cambridge, MA: MIT Press; Tyson, L (1992) *Who's Bashing Whom: Trade Conflict in High-Technology Industries*, Washington DC: International Institute for Economics.
14. Tyson, 1992:14.
15. *Ibid*:13.
16. Florida, R & Browdy, D (1991) 'The Invention that Got Away', *Technology Review*, August/September, pp. 43–54.

2
National Systems of Innovation and Path Dependency

It was in the late 1980s when the *systems of innovation* concept first came to major prominence. When Christopher Freeman's (1987) book on innovation in Japan was published,[1] where the concept in its original formulation made its initial appearance as the national system of innovation (NSI), no one could have predicted the effect which it would have. The main effect Freeman had was to help create an entirely new field of research and spark off a wave of research into the concept. Within six years two major books had appeared on the subject, *National Systems of Innovation: Towards a Theory of Innovation and Interactive Learning* (1992) edited by Bengt-Åke Lundvall, and *National Innovation Systems: A Comparative Analysis* (1993) edited by Richard Nelson. Since then, the systems of innovation concept has continued to attract a considerable degree of interest for two fundamental reasons; it has proved highly robust and adaptable; and in an era of globalisation it has generated important insights across a variety of different contexts within the process of innovation (Michie, 2003).[2]

A measure of the systems of innovation concept's general importance and influence can be found in Lundvall *et al.* (2002) who point to the fact that international organisations such as the OECD, the European Commission, and UNCTAD have absorbed it as an integral part of their analytical perspective. Countries such as the US and Sweden have given it their 'official' approval in very different ways; the US Academy of Science has brought into its vocabulary the national innovation system and now uses it as a framework for analysing science and technology policy in the US; and Sweden has named a new central government institution VINNOVA, (an 'ämbetsverk') which stands for 'the Systems of Innovation

Authority'.[3] Lundvall *et al*. add that interest in the NSI perspective is
now growing in Asia and Latin America.[4] It is no wonder then
Nelson (2002) describes the systems of innovation concept 'as an
institutional concept, par excellence'.[5]

The extent of the concept's diffusion and influence is a reflection not
only of its flexibility, but also of its value as a major analytical tool that
can be effectively applied across an immense range of problems and
issues. It is this versatility that makes it so useful as one of the main
tools to analyse a highly important and complex situation concerning
the process of industrial catch-up, which has been completely reversed
in semiconductors and LCDs. In virtually every other industry,
whether it has been in steel, shipbuilding, or machine tools, it is the
newly industrialising countries (NICs) that have had to catch up with
the advanced industrialised countries (AICs) of the West. However,
TFT-LCD is an example of an industry where the tables have been
turned on the AICs. South Korea and Taiwan have been the only countries to successfully challenge Japan in TFT-LCDs. Semiconductors and
TFT-LCDs are two technologies of a global nature that have seen the
'pendulum' of innovation swing more recently in the favour of East
Asian firms and away from Western firms. This is obviously a very
important development, and could possibly herald the beginning of a
new era where Western firms become increasingly reliant on East Asian
firms for technology rather than vice-versa. It is East Asian firms that
are in the ascendancy, and furthermore, they seem to have a much
greater appreciation of the cumulative nature of technological change
than Western firms do.

The most obvious catalyst for the enormous growth in the interest in
the systems of innovation concept was Japan's success in a host of
high-profile industries since the late 1970s. As Japan's expansion into
overseas markets gathered pace, the inevitable questions started to be
asked as to what exactly lay behind its success. In essence, this was no
different from the questions that were being asked about what were
the causes of the US economic malaise during the same period. What
differentiated Japan from the US, however, was that its success was
more often than not in areas of rapid technological change such as
electronics. This made the task of identifying the sources of its 'success'
that much more difficult than the sources of US 'decline'. Nelson
(1993) points to this growing divergence between nations, and the
need to understand the similarities and differences, and the extent to
and manner in which these differences explain variation in national
economic performance.[6] What was particularly striking at this point

was the rapidly changing position of the US; it seemed far less enviable than it had once been. As Nelson (1993) points out:

> Until the 1970s there was no strong competition to the American system as a broad model of how an innovation system should be designed. This standing as a model was a natural reflection of the US technological pre-eminence that marked the post war years...As European productivity and income levels have caught up with American levels, and Japan has emerged as a leading economic and technological power, the attraction of the American model has waned and Japanese institutions have waxed as targets for emulation.[7]

At the same time the rapid growth of Japan was putting immense pressure on the US model, to use Nelson's own term, it was the rapid emergence of two other nations, South Korea and Taiwan, which seemed to sound its death knell. In retrospect it would be easy to conclude that the prevailing pessimism over the relative position of the US was overdone, however, there were good reasons for it. Concerns about the country's economic and technological performance could be traced back to the 1970s. Problems such as declining productivity and US firms' lack of willingness to innovate were seized upon by Hayes & Abernathy (1980) in their classic paper, 'Managing Our Way to Economic Decline'. They contend the root of these problems was US firms' 'chronic' short-termism which attached far too much emphasis on the importance of financial objectives at the 'expense' of innovation. This 'management by numbers' had become so deeply embedded and widespread the fear was unless something was done the US risked surrendering its technological leadership to its competitors, namely Japan, for good. As it turned out the 1990s were to witness another remarkable change – a reversal in the change in the fortunes of the US and Japan.

The once pre-eminent model, the US NSI, was given a new lease of life and its Japanese counterpart fell from grace almost as fast as it had arisen. In part, this turn of events was affected by the change in the prevailing economic climate of the respective countries. Following the end of the 'bubble' years, the second half of the 1980s when Japan experienced a new period of high growth peaking at 5.6 per cent in 1990, the country then fell into a period of prolonged economic stagnation which lasted the best part of the decade. Compare this to the US, which experienced a general economic recovery in the early 1990s, and then on the back of this went on to enjoy a spectacular technology boom in the mid to late 1990s. The recovering fortunes of the US led Berggren & Nomura (1997) to observe: 'The US resurgence is reflected in an increasing lead in high-tech sectors such as biotechnology, computers and software, as well as in dramatic turnaround in traditional manufacturing sectors which were under heavy siege in the 1980s'.[8] As Berggren & Nomura show, this change at the 'top' led to sharply divergent views on the extent of the US 'comeback' and Japan's 'stumble'. This debate proved in many ways very unhelpful. Apart from clouding the true situation between the two countries, it also obscured the wider picture.

Unlike the 1980s where the focus was almost exclusively on the US and Japan, the 1990s saw much of the focus switch away and towards South Korea and Taiwan. As stated above, their rapid emergence put enormous pressure on the US and Japan, especially in sectors where they had enjoyed unrivalled technological dominance, semiconductors being a prime example. The growth and success of South Korea and Taiwan dispelled the notion that the US NSI or Japan's NSI were somehow 'superior' to other NSIs, and they could no longer be realistically touted as models other countries should seek to emulate. These developments inevitably prompted questions about what were the factors that lay behind the success of South Korea and Taiwan in highly dynamic and competitive sectors, and how was it that Western Europe could not even begin to match them? Potential answers lay in a closer analysis of these countries' NSIs.

Origins of the concept

Odd as it may seem, the concept is not a modern one by any means. According to Freeman (1997) the concept can be traced back to at least Friedrich List (1841) and his notion of 'The National System of Political Economy'.[9] In his illuminating account Freeman argues that List could equally have called it 'The National System of Innovation', as he was highly aware of the many factors which are important to any modern industrial economy, for example, investment, institutions, the import of foreign technology, education and training, and the numerous links which exist between them.[10] The prime concern of List was how Germany would overcome its economic backwardness relative to England, then the world's leading industrial nation, and how it would catch up and overtake England. As Freeman explains, in the strategy List mapped out for Germany 'he advocated not only protection of infant industries but a broad range of policies designed to accelerate or make possible, industrialisation and economic growth. Most of these policies were concerned with learning about new technology and applying it'.[11] The single most prominent feature of the strategy was the *proactive* role of the state. List realised the interdependence between the importation of technology and domestic technical development[12] and concluded that for Germany to develop and build up its own technological capabilities, the state would need to coordinate (and implement) long-term policies formulated for this very purpose.

List's work was given new credence when it was found to be of great significance to the contemporary world. When Freeman's (1988) analysis of Japan's NSI is compared with what List advocated for Germany in the nineteenth century, the parallels between them are striking. The two factors that stand out beyond anything else are the overall long-term effectiveness of policies aimed at the development of the technological capabilities of each country, and the proactive role of the state. When analysing the role of the state in Japan's post-war economic success, the usual focus has been on the role of the Ministry of International Trade & Industry (MITI), about which a great deal has been already been written. Freeman, however, points to a number of its roles which are generally not appreciated, but have nevertheless proved extremely valuable. Amongst others, these include its long established system of technological forecasting, the promotion of the most advanced technologies with the widest world market potential over the long term, and helping to provide the necessary infrastructural investments.[13] The other aspects of Japan's NSI which Freeman touches upon are MITI's long-term relationships with Japanese firms and its initiatives to promote interfirm collaboration, which have been crucial for the effective implementation of MITI's policies, and the importance of general education and training.[14] Needless to say, the role that the state has played in Germany and Japan laid the basis for their subsequent economic success. In both cases, it is what the state *did* and *how* it did it, which is the crucial aspect. The long-term success of these policies is still evident today.

Definition and evolution of the concept

In light of the considerable amount of research which has been undertaken and the general popularity of the concept, it would be reasonable to assume that there was a 'standard' definition for the NSI, similar to the one that exists for the term 'innovation'.[15] Surprisingly though, no such definition exists. Much of this has to do with the complexity of the innovation system and its various elements. As Lundvall (1992) has explained, it is possible to have both a 'narrow' and 'broad' definition of the NSI, depending on what is included and what left out.[16] The NSI has a number of very different elements which all interact with each other in one way or another. By and large, the NSI consists of the internal organisation of the firm, interfirm relationships, the role of the public sector, the institutional arrangement of the financial sector, R&D intensity and R&D organisation.[17] With such an array of different elements it is little wonder that a 'standard' definition has proved so elusive. In an attempt to bring greater clarity to the concept, Niosi *et al.* (1993) have developed a 'workable' definition, which is given below:

> A national system of innovation is the system of interacting private and public firms (either large or small), universities and government agencies aiming at the production of science and technology within national borders. Interaction among these units may be technical, commercial, legal, social and financial, inasmuch as the goal of the interaction is the development, protection, financing, or regulation of new science and technology.[18]

More recently, Galli & Teubal (1997) have defined the NSI as 'the set of organisations, institutions, and linkages for the generation, diffusion, and application of scientific and technological knowledge, operating in a specific country',[19] which, as they demonstrate in their analysis; has, many complex inter-linkages. At the very least these definitions confirm Lundvall's earlier point that the NSI is a complex animal and is made up of a number of different elements. Exactly how 'useful' these individual definitions are depends on the perspective from which they are viewed. Niosi *et al.* maintain the dominant element in most NSIs is the state, on the grounds that it finances and carries out a large share of national R&D.[20] They point to various countries, for example, the US and France, where the state accounts for 49 per cent and 50 per cent, respectively, of national R&D.[21] While the state does have a pivotal role in the functioning of an NSI, it is the firm that determines its overall 'efficiency'. The state can organise all kinds of research initiatives and programmes to stimulate innovation, but the firm is ultimately responsible for whether an innovation succeeds or fails.

If the definitional problems of the concept were not bad enough, globalisation has made it worse. In seeking to define the NSI, the impression given is that a country's innovation activities and processes are *confined* to national borders, when nothing could be further from the truth. One of the unmistakable trends of globalisation has been the huge growth in the proliferation of technological alliances and international joint ventures between firms, for example, multinational corporations (MNCs). The main effect of globalisation on NSIs has been to make them more open systems (Galli & Teubal, 1997). The key question that remains is how globalisation has affected NSIs in the development of specific technologies, which is an issue that is returned to later.

It was stated earlier that a reason why the concept of innovation systems has become so influential is that it has proved highly adaptable. Freeman's original concept has spawned the development of three other major related concepts that still emphasise the systemic characteristics of innovation. The first of these concepts is *technological systems* which was developed in the mid-1990s by Bo Carlsson. Carlsson has defined technological systems 'as a network or networks of agents interacting in a specific technology area under particular institutional infrastructure to generate, diffuse and utilise technology'. Some of the technologies which have been analysed using this concept include factory automation, materials technology and pharmaceuticals (Edquist, 1997).[22]

The second concept is *regional systems of innovation*, which like the NSI concept before it has attracted considerable attention (Dalum *et al.*, 1999, Howells, 1999, Mytelka, 2000, Cantwell & Iammarino, 2003).[23] It has been defined as 'the localised network of actors and institutions in the public and private sectors whose activities and interactions generate, import, modify and diffuse new technologies' (Cantwell & Iammarino, 2003). This concept has proved especially fruitful when compared to technological systems. In the context of globalisation it has become increasingly necessary to understand the importance of location (Cantwell & Iammarino, 2003); why some regions such as California's Silicon Valley are far more innovative and dynamic than others; what the similarities and differences are between regions; and how the regional system of innovation interacts with the wider NSI. In essence, the concept seeks to explain how and to what extent the institutional and cultural environment of a region supports or obstructs innovation (Kaiser & Prange, 2004).[24]

The third and final concept is *sectoral systems of innovation* which Freeman & Soete (1997) have described as one of the most exciting fields for the new millennium. The grounds for this excitement are straightforward. The innovative environment varies immensely across sectors. Malerba (2004) observes that in certain industries large firms have a dominant role and are responsible for a high percentage of innovations, for example, in the chemical industry. The dominance they enjoy in this industry is not replicated in other industries such as telecommunications and software.[25] Malerba analyses a range of sectors; pharmaceuticals, chemicals, Internet and mobile communications, software, machine tools, and services. He shows that Europe's performance across these sectors is 'mixed'. Here, the same approach is used to analyse two high technology industries at a global level without a specific European focus.

Broadly speaking, the sectoral system of innovation can be defined as 'a set of new and established products for specific uses and the set of agents carrying out market and non-market interactions for the creation, production, and sale of those products. Sectoral systems have a knowledge base, technologies, inputs, and an existing, emergent and potential demand'.[26] Although the definition given by Malerba is more comprehensive than the one given here, it shares a characteristic with defining the NSI in that it can be made broader or narrower depending on the research question, rather than on what elements of the system are included and what is left out. This may seem relatively unimportant, but Malerba makes a number of pertinent points about sectors that are often overlooked. First and foremost, sectors undergo a process of change and transformation over time. Secondly, there are enormous variations between sectors in terms of size, composition, firm strategies and organisation, the rate and direction of technological change, and networks among actors. In the current context, the advantage of this particular approach is that it allows important insights to be generated into how specific high technology industries have been transformed over time. Also, more specifically, it allows a focus on elements of change such as firms' strategies and firm learning. This will provide some initial answers to some of the major questions Malerba raises about how do new sectoral systems emerge, and what is the link with previous sectoral systems?

These different concepts enable the pioneering work of Nelson (1993) and Lundvall (1992) to be built upon and extended. Nelson analysed the NSIs of fifteen countries to examine the similarities and differences between their institutional set-up, which is structured to support, shape, and advance technological innovation. The differences found between these systems centred around such aspects as university research, industrial R&D, public infrastructure, financial institutions, training and education, and management styles. In his treatment of the NSI, Lundvall (1992) emphasised the central importance of learning. Lundvall's purpose for placing such an emphasis on learning, as pointed out by Archibugi *et al.* (1999) is that it serves a dual purpose; as a key element in both the *dynamic* of the system and as a key agent in *binding* the whole system together.[27]

These pioneering studies collectively highlighted how extensive the differences which exist between countries are, and the specific elements that are critical to a country's NSI overall effectiveness. However, in neither of these studies was any comparative analysis included to illustrate how the differences between NSIs can affect the performance of various countries in a major industry. This is an important gap in the current literature. Sufficient time has elapsed since the publication of these major studies, affording the opportunity to fill this void. Furthermore, the sectoral systems concept can be coupled with the NSI to analyse why some NSIs are better at supporting newly emerging sectors than others.

Comparing the performances of NSIs is not a straightforward process. The methods most commonly used to compare the performances of different NSIs are R&D expenditures and output indicators such as patents, the proportion of high technology products in foreign trade, and the proportion of new products.[28] But as Lundvall (1992) notes, neither of these approaches has proved very satisfactory. The two major problems he has identified with R&D expenditures (as a proportion of GDP) are that they do not give any indication of what the end product will be, and secondly, they are only one of the many aspects that make up the innovation process.[29] Each of the output indicators is flawed.[30] Furthermore, as a group they are unable to accommodate one of the most important elements of the innovation process, learning.[31] With inherent weaknesses plaguing the two most commonly used methods, there is a need to find much more comprehensive methods for comparing NSIs. As the task in hand is so complex there is, perhaps not surprisingly, no consensus on what is the best method to use for making accurate and reliable comparisons.[32]

A more appropriate method is to focus at the sectoral level. The advantage of this approach is that the efficiency and health of any NSI in a particular category can be easily measured through fluctuations in market shares. Although, like its predecessors it is far from being perfect, it does have the added advantage of being able to be used in conjunction with Patel & Pavitt's (1994) system of classifying NSIs.[33] They contend that a useful first step when making comparisons between NSIs is to distinguish between those that are 'myopic' and 'dynamic'.[34]

What sets the two systems apart from each other is how they treat technological activities.[35] In 'myopic' systems, investments in technological activities are treated the same as conventional investments. When decisions are made about conventional investments they are based predominantly on short-term considerations (Hayes & Abernathy, 1980). The main criterion used to evaluate projects is rate of return on investment, the pay back period that is typically around three years, and the maturity of the market. Because of their very nature, investment in technological activities involves accepting a high degree of risk and uncertainty. However, when they are evaluated using the same criteria as conventional investments there is a great tendency for them to appear 'unattractive'. The type of NSI structure make them *passive* learning systems.[36] In 'dynamic' systems, investments in technological activities are treated completely differently from conventional investments.[37] When investments are made in technological activities, they are made with the long term very much in mind. Risk, market uncertainty, and high levels of investment are all seen as being the norm. This makes it much easier to follow certain technological trajectories than would otherwise be the case. Investments made in early generations of a technology are often rapidly followed by subsequent programmes of investment as the technology starts to mature. This insures the process of learning goes uninterrupted and helps develop vital technological competencies.[38] The dynamic systems are structured as *active* learning systems.[39] The linkage between investment and learning would appear to be the 'spinal cord' of the NSI, and should it become damaged in any way, its capacity to maintain existing and develop new technological capabilities (in old and new sectors) will be seriously impaired.

Globalisation and the NSI

In the era of globalisation there has been an ongoing debate between two key commentators in particular, Michael Porter and Kenichi Ohmae, who have competing views over the continued importance of a firm's national base. According to the OECD (1992), globalisation in this context refers to an evolving pattern of cross-border activities of firms involving international investment, trade and collaboration for purposes of product development, production, sourcing and marketing. These international activities enable firms to enter new markets, exploit their technological and organisational advantages and reduce business costs and risks. Underlying the international expansion of firms, and in part driven by it, are technological advances, the liberalisation of markets and increased mobility of production factors.[40]

In *The Borderless World*, Ohmae (1990) argues that national borders are on the brink of becoming permanently obsolete in what he refers to as the ILE (interlinked economy) of the USA, Japan, Western Europe, and the newly industrialising countries (NICs). He goes far as to state: 'It [the ILE] is becoming so powerful that it has swallowed most consumers and corporations, made traditional national borders almost disappear, and pushed bureaucrats, politicians and the military toward the status of declining industries'.[41] Elsewhere, he remarks, 'When money goods, people, information and even companies crisscross national borders so freely, it makes no sense to talk of "American industrial competitiveness"'.[42] Ohmae accepts that prior to the 1980s national policies and national differences were important.[43] Compare this to Porter's (1990) tone:

> Competitive advantage is created and sustained through a highly localised process. Differences in national economic structures, values, cultures and histories contribute profoundly to competitive success. The role of the nation state seems to be as strong or stronger than ever. While globalisation of competition might appear to make the national less important, instead it seems to make it more so. With fewer impediments to trade to shelter uncompetitive domestic firms and industries, the home nation takes on a growing significance because it is the source of skills and technology that underpin competitive advantage. [44]

These two radically different perspectives have created a confusing picture to the point where it is impossible to distinguish accurately who is correct? At a broad level, they are both correct in various respects. This requires some careful quantification, however. The broad thrust of Ohmae's analysis is how MNCs have become more consumer-orientated with the rise of consumer 'power' and the need to serve the global marketplace. He is correct to identify the proliferation of strategic alliances between firms from different continents. Where his analysis is wide of the mark is in his treatment of technology. He neglects how new technologies can emerge and firms' strategies to develop them. New technologies are 'out there' at the disposal of firms. His supposition is that technologies become widely available because the effective control of technology is very difficult and individual firms cannot afford to develop every critical technology. This lack of any serious discussion of the subject leads Ohmae not to give the role of new technologies any special prominence. Porter's assessment, on the other hand, is more exact. With globalisation showing no signs of abating, the debate between Ohmae and Porter inevitably raises questions about the whole concept of the NSI up to the point of making it irrelevant.[45] Porter's analysis and the empirical evidence suggests that while globalisation has changed the nature of the NSI and will continue to do so there are very few signs that it will become an anachronism soon.

NSI and firm strategy

The 'correlation' between the NSI and firm strategy is far from clear-cut. Nevertheless, the experiences of the semiconductor and LCD industries suggest that where the NSI has the *capacity* to support a highly capital-intensive and dynamic technology, there is a very high probability that a firm's strategy will be focused on the development of the technology in question. The semiconductor industry is a case in point: since the 1950s a number of different countries have enjoyed unrivalled technological leadership of the industry, before then being overtaken by a more 'nimble' competitor. From the 1950s to the beginning of the 1980s the USA enjoyed an unrivalled dominance of the industry, before it was subsequently overtaken (in DRAMs) by Japan. In the late 1980s and early 1990s Japan's stranglehold on the semiconductor industry was broken by South Korea, which has subsequently established a major competitive advantage in DRAMs. Prior to the mid-1990s much attention had been focused primarily on the DRAM industry and for good reason. These developments have now been overshadowed by new and important trends which have in part been brought about by the changing economics of the industry and the astonishing rate of technological change that the industry has been experiencing.

The first of these new developments has been the rapid emergence of Taiwan and its foundry 'model' for producing semiconductors. The speed of its emergence has been almost as rapid as that of South Korea. Until the mid-1990s the manufacture of semiconductors had been the preserve of large vertically integrated firms, but Taiwan has developed a new 'made to order' system for firms wanting to produce semiconductors. As the price of new fabrication plants (or fabs) has now reached stratospheric levels, fewer and fewer firms have found themselves able to afford the kind of levels of investment required and have become much more willing to outsource the manufacture of semiconductors on a 'made to order' basis. Taiwan is now the undisputed leader in this field, just as South Korea is in DRAMs. In both cases, the NSI has been of critical importance in enabling the respective firms of these countries to become highly competitive at a global level. These developments have now been overshadowed by new and important trends which have in part been brought about by the changing economics of the industry and the astonishing rate of technological change the industry has been experiencing.

The second of these trends has been the industry-wide rush to develop increasingly sophisticated and complex semiconductors. Faced with a severe slump in demand and falling prices in 2001–02, the development of a new generation of semiconductors offered firms a way of improving their long-term profitability. Because of cutthroat competition, especially in DRAMs, two of the main areas that firms are developing are MRAM (magnetic random access memory) chips and Nand flash memory. The latter are used in digital devices such as personal computers, mobile phones, and games consoles. In conjunction with the development of this new generation of semiconductors, the industry has also witnessed an unprecedented wave of strategic alliances between firms that would have been unthinkable a decade ago. Alliances have been formed, or are in the process of being formed between Japanese firms, Taiwanese and European firms, and Japanese and US firms. With the emergence of Taiwan on the one hand, and the growth of strategic alliances on the other, it demonstrates that in a truly global industry the NSI still has a crucial role to play, whilst at the same time also highlighting its limitations.

From a strategic perspective, one of the most interesting aspects of the semiconductor industry in the current context has been the pivotal role US firms have always played. As mentioned earlier, for the best part of three decades US firms enjoyed an unrivalled dominance of the industry before it was successfully challenged by Japan in the early 1980s. Since then, there has been a perceptible and fundamental shift in the strategies of many US firms, which became especially pronounced in the 1990s, that is a switch away from the production of semiconductors to their design. Whether it is in DRAM technology, foundries, or the use of nanotechnology in the production of semiconductors, US firms are no longer the 'pioneering' firms in the industry, in other words, those that are prepared to lead from the front. This might have something to do with the fact that US firms have become increasingly 'fabless'. They specialise in the design of semiconductors, but do not have any production facilities of their own, relying on (Taiwanese) foundries for the production of their designs.

Hayes & Abernathy (1980) offer a convincing and plausible explanation of how this came about. The pursuit of short-term financial objectives has witnessed the rise of a new management orthodoxy known as 'pseudo-professionalism'. These are large numbers of individual managers with little knowledge or expertise of an industry or technology who transfer from one firm to another after a short period, having 'turned it around', a euphemism for restoring profitability. A manager's individual success is commonly measured by the level of profits that can be generated over the short term, a period often lasting no more than three years. One of the most common approaches used to generate such profits has been to use the rate of return on investment (ROI) method of analysis. The long-term costs of this approach and others of a similar type are clear for everyone to see. Projects that are risky and may not yield a profit have little chance of ever getting off the drawing board. And at the same time these are the very same projects on which many firms' long-term future depends for they are the source of new technologies, products and processes. In sum, managers who believe it is more important to 'manage by numbers' are seriously misguided.

Perhaps without realising it, what they are doing is undermining their own firm's long-term competitiveness and putting their long-term survival in jeopardy. How can a firm retain its competitiveness by being 'averse to risk' within an industry where the threat of technological change is ever present and its rivals are constantly looking for ways to steal a lead? The obvious answer is that it cannot. Scherer (1992) found that 'poor management', triggered by the Hayes & Abernathy 'syndrome' in the 1980s was responsible for the US losing its technological lead across a diverse range of industries, such as machine tools, automobile tyres and facsimile machines, and the loss of significant market share in both its domestic and overseas markets.[46]

The Hayes & Abernathy syndrome hit the US semiconductor industry in the mid-1980s. During the industry recession (1985–87) a large number of US firms left the DRAM industry in the face of intense Japanese competition and falling prices. Once demand and prices had recovered many of the firms were never to return. For the sake of short-term profitability US firms effectively 'abandoned' the technology in complete contrast to Japanese and South Korean firms, which continued to pour investment into DRAMs. In the recession

Japanese firms suffered financially in the short term as much as any US firm did. Japanese firms though, unlike their American counterparts showed a much greater appreciation of the strategic importance of DRAM technology – it was one of their core competencies. As Prahalad & Hamel (1994) point out, the problem with core competencies is that they need to be continuously nurtured and protected, otherwise they will seriously impair a firm's future competitiveness in new growth areas of business. They go on to say that Western firms hardly think of competitiveness in these terms and have shown far less inclination to preserve their own core competencies in the same fashion as Japanese firms have. Sony is a classic example of a Japanese firm that strives to continuously improve its core competencies in miniaturisation, whilst Motorola is one of several major Western firms that have fallen into the trap of allowing its core competencies to 'decline'. When Motorola, for instance, decided to skip one generation of DRAMs and not develop the 256K generation, it required considerable technical assistance from Japanese firms to compete in the following generation, the 1M DRAM.

Undoubtedly, the most damaging long-term effect of the Hayes & Abernathy's syndrome for US firms has been the very serious erosion of their core competencies in DRAM technology. When a new industry emerged in the early 1990s, LCDs, which has very close links with DRAMs, US firms naturally wanted to enter the industry because of its enormous growth potential. Time was to show that their core competencies had literally been destroyed. By the end of the 1990s it had become obvious that US firms were incapable of competing in LCDs. As with semiconductors, the US was the first country to establish a technological lead in LCDs, largely because of pioneering work carried out by two highly innovative firms, Westinghouse and RCA, in the late 1950s and early 1960s. This time though it failed to take advantage of this lead. Due to the Hayes & Abernathy syndrome, the US let the technology slip through its fingers. Again, it is the NSI that is key to understanding why this very unusual situation has arisen.

In the foregoing discussion, it can be seen that the NSI concept is very useful for comparing the performances of different countries in a specific technology at the sectoral level. Two of its major 'limitations', however, are that it cannot be used to analyse the evolution of technologies, nor changes at the level of the firm, which by implication are immensely important. To do this, it is necessary to supplement the NSI concept with an additional perspective that can provide a much fuller account.

Path dependency

Path dependency is now seen as an <u>interdisciplinary</u> concept and has <u>been used as an analytical tool in diverse fields to analyse a wide range</u> <u>of different phenomena (Hirsch & Gillespie, 2001).</u>[47] Nevertheless, the one idea that has become intrinsically linked with it is that 'history matters'. With its origins firmly rooted in the studies of David (1985) and Arthur (1989)[48] the advantage of the concept is that it is possible to chart the importance of the sequence of specific micro-level historical events (Ruttan, 2001),[49] especially the role of the firm. As David (1986) states:

> it is sometimes not possible to uncover the logic (or illogic) of the world around us except by understanding how it got that way. A *path-dependent* sequence of economic changes is one in which important influences upon the eventual outcome can be exerted by temporally remote events, including happenings dominated by chance elements rather than systematic forces. Stochastic processes like that do not converge automatically to a fixed-point distribution of outcomes, and are called *non-ergodic*. In such circumstances 'historical accidents' can neither be ignored, nor neatly quarantined for the purpose of economic analysis; the dynamic process itself takes on an *essentially historical* process.[50]

In the above passage, David infers that on certain rare occasions the best way to solve a conundrum, that is, to understand an 'unusual outcome' is to carry out an analysis of (unspecified) economic changes. With this approach, David postulates that it is feasible to establish which changes have been the most influential and why. This is all very well and good up to a point. David 'misses' one important point, however. A distinct inference is made that when analysing any 'unusual outcome' there is an 'obvious' beginning and end, as is the case with most 'paths'. For any analysis of this nature to be as comprehensive as possible, it is important to know where the path begins. In addition, there is also the need to widen the scope beyond economic changes as this focus is too narrow for purposes of analysis. In the current context, the path dependent approach will take the form of an analysis which examines a myriad of changes (some of which are not economic) and places them in the appropriate chronological order. This makes the task of examining those changes which have had the greatest effect that much easier. Here, the non-ergodic characteristic of a process can really come into its own; as David stresses, systems which possess this property cannot shake off the effects of past events.[51] This is certainly true for LCDs.

In the case of LCDs the path begins somewhere between the 1850s and 1880s when scientists in Europe carried out a series of chemical experiments which eventually led to the 'discovery' of liquid crystals in 1888. The path description starts in the nineteenth century and not at the beginning of the 1970s (when LCDs first appeared on the market) because of one key point that Rosenberg (1994) makes about the cumulative nature of technological change.[52]

Rosenberg points out that the trajectories that technological change proceeds down are often heavily reliant upon the stock of previous technological (and scientific) knowledge. He highlights the fact that it is of no real benefit to simply 'make reference to certain initial conditions'[53] when attempting to understand the development of a major innovation such as the electric-power plant, the transistor, and computer.[54] These three innovations have provided the foundations for other innovations, some of which are directly linked to them, and others which are not. Rosenberg goes on to say that to understand why technological knowledge has taken the direction it has in a particular case, and where it is likely to go in the future, 'can only be understood within the context of the particular sequence of events, which constitutes the history of the system'.[55]

To illustrate his point, Rosenberg highlights the curious 'link' between electric power and the development of the semiconductor industry. He claims it is highly improbable whether the invention of the vacuum tube and the transistor would have been possible without some prior development of some sort of electric-power generating capability.[56] It is important to keep in mind that the development of the vacuum tube laid the basis for the invention of the transistor. What is not widely known is that the transistor was the outcome of a research programme carried out by AT&T to reduce its heavy reliance on vacuum tubes (which were not cost-effective) for long lines switching.[57] In turn, the transistor laid the basis for the development of the integrated circuit (IC), which itself stimulated the development of completely new methods of production for semiconductors, and a huge growth in chemical science.[58] The main point that Rosenberg is trying to make is that 'one thing follows another', but as is shown in the above example, the sequence of events is sometimes far from obvious, and this is precisely the case with LCDs.

As mentioned earlier, the chemical experiments which were carried out in Europe between the 1850s and 1880s and led to the discovery of liquid crystals in 1888, effectively laid the foundations for the modern LCD industry. In Chapter 6, it is stated that this important breakthrough led to a surge of interest in this new line of scientific enquiry. This high level of interest in liquid crystals lasted until the beginning of the twentieth century, before dropping off quite substantially until the 1920s, when there appears to have been a resurgence of interest which lasted until the war, with pioneering work being done in a number of different countries. After the war interest in liquid crystals once again fell away, before experiencing yet another resurgence in the late 1950s. From the brief account above, it is clear to see that the path which the development of liquid crystals took up to this point is not only easy to follow, but more importantly, there is nothing out of the ordinary about it. Suffice to say, that the path remains easy to follow until the beginning of the 1970s when the LCD made its first appearance on the market. After this date the path starts to 'twist and turn', making analysis that much more difficult. However, it is made all the easier by employing the ideas of Arthur (1989) about competing technologies.[59]

The essence of Arthur's argument is that when two (or more) technologies compete for a market there are two possible outcomes. The first is when an inferior technology locks out other technologies because it is the first to appear on the market by chance, which then later goes on to 'corner the market'[60] of potential adopters.[61] A widely cited example of this is the QWERTY keyboard (David, 1986). The second potential outcome is when a technology becomes dominant because of what Arthur terms 'insignificant events', such as the whims of early developers and the unexpected successes in the performances of prototypes.[62] These, in due course, permit the technology concerned to achieve sufficient adoption and improvement for it to eventually become dominant.[63]

When it comes to trying to understand why the LCD industry experienced 'take off' in the early 1970s, the second outcome, which Arthur discusses, allows everything to be put into its proper context. In the manufacture of digital watches there were two main display technologies, the LCD and light-emitting diode (LED). It was at this stage when the LCD started to 'compete' with the LED. The various improvements which the LCD underwent allowed it to achieve sufficient adoption during the course of the 1970s to subsequently mount a successful challenge to the LED, within only a few years of appearing on the market. In Chapter 6, we shall see that it was the whims of the early developers who believed the LCD had great potential, and the unexpected successes of the early prototypes which gave the LCD the foundations it needed in order to be able to do this. In addition, the importance of specific 'historical accidents' will be shown to be vital to our understanding of the evolution of the industry.

All that remains to be said here is that Rosenberg's point that the trajectories that technological change proceeds down are often heavily reliant upon the stock of technological (and scientific knowledge) could not be more accurate when it comes to the trajectory the LCD industry has followed. To see how important the stock of knowledge about liquid crystals has been to bring the LCD in its most basic form to the marketplace, one only has to look at the time gap between when liquid crystals were discovered and when the first LCD appeared on the market. The time gap was approximately ninety years. Since the 1970s, not only has the stock of knowledge about liquid crystals experienced phenomenal growth, it has also been necessary to expand and draw upon another stock of knowledge on thin-film transistors (TFTs). This has been absolutely essential if the TFT-LCD was to ever reach the market place.

Summary

This chapter has analysed the importance and flexibility of the systems of innovation concept, paying particular attention to the continued importance of the NSI in an era of globalisation. At the same time, it has been argued that the NSI can be coupled with the sectoral system of innovation concept as the starting point to analyse the wide variations in the performance of different countries in TFT-LCDs, using the semiconductor industry as the key 'template'. Furthermore, path dependency can be used as a supplementary framework as a means of tracing a sequence of events at the level of the firm to understand the wider dynamics of the evolution of the LCD industry.

Notes

1. Freeman, C (1987) *Technology Policy and Economic Policy: Lessons from Japan*, London: Pinter.
2. Michie, J (2003) (ed.) *The Handbook of Globalisation*, Cheltenham: Edward Elgar.

3. Lundvall, BA, Johnson, B, Sloth Anderson, E & Dalum, B (2002) 'National systems of production, innovation and competence building', *Research Policy*, Vol. 31, No. 2, p. 214.
4. *Ibid*:216.
5. Nelson, RR & Nelson, K (2002) 'Technology, institutions, and innovation systems', *Research Policy*, Vol. 31, No. 2, p. 265.
6. Nelson, R (ed.) (1993) *National Innovation System: A Comparative Analysis*, New York: Oxford University Press, p. 3.
7. *Ibid*:16–17.
8. Berggren, C & Nomura, M (1997) *The Resilience of Corporate Japan: New Competitive Strategies and Personnel Practices*, London: Paul Chapman, p. 23.
9. Freeman, C & Soete, L (1997) *The Economics of Industrial Innovation*, London: Pinter, p. 295.
10. *Ibid*.
11. *Ibid*.
12. *Ibid*:297.
13. Freeman, C (1988) 'Japan: a new national system of innovation?' in G. Dosi *et al.* (ed.) *Technical Change and Economic Theory*, London: Pinter, pp. 331–334.
14. *Ibid*:334–346.
15. The generally accepted definition for innovation is the one which has been developed by Freeman.
16. Lundvall, BA (1992) *National Systems of Innovation: Towards a Theory of Innovation and Interactive Learning*, London: Pinter, p. 12.
17. *Ibid*:13.
18. Niosi, J, Saviotti, P, Bellon, B & Crow, M (1993) 'National Systems of Innovation: In Search of a Workable Concept', *Technology in Society*, Vol. 15, No. 2, p. 212.
19. Galli, R & Teubal, M (1997) 'Paradigmatic Shifts in National Innovation Systems' in C. Edquist (ed.) *Systems of Innovation; Technologies, Institutions and Organisations'*, London: Pinter, p. 345.
20. Niosi *et al.* 1993:212.
21. *Ibid*. The figures given by Niosi are for 1988.
22. Edquist, C (1997) 'Systems of Innovation Approaches – Their Emergence and Characteristics', in C. Edquist (ed.) *Systems of Innovation; Technologies, Institutions and Organisations'*, London: Pinter, p. 5.
23. The sources referred to are: Howells, J (1999) 'Regional Systems of Innovation' in D Archibugi, J Howells & J Michie (eds) *Innovation Policy in a Global Economy*, Cambridge: Cambridge University Press; Dalum, B, Holmen, M, Jacobssen, S, Preast, M, Rickne, A & Villumsen, G (1999) 'Changing the regional system of innovation' in J Fagerberg, P Guerrieri & B Verspagen (eds) *The Economic Challenge for Europe*, Cheltenham: Edward Elgar, pp. 175–200; Mytelka, L (2000) 'Localised Systems of Innovation in a Globalised World Economy', *Industry and Innovation*, 7, pp. 15–32; Cantwell, J & Iammarino, S (2003) *Multinational Corporations and European Regional Systems of Innovation*, London: Routledge.
24. Kaiser, R & Prange, H (2004) 'The Reconfiguration of National Innovation Systems – the example of German biotechnology, *Research Policy*, Vol. 33, No. 3, pp. 395–408.

25. Malerba, F (ed.) (2004) *Sectoral Systems of Innovation: Concepts, Issues and Analysis of Six Major Sectors in Europe*, Cambridge: Cambridge University Press.
26. Malerba, F (2002) 'Sectoral Systems of Innovation and Production', *Research Policy*, Vol. 31, No. 2, p. 250.
27. Archibugi D, Michie, J & Howells, J (eds) (1999) *Innovation Policy in a Global Economy*, Cambridge: Cambridge University Press, p. 5. Authors' italics.
28. Lundvall, BA (ed.) (1992) *National Systems of Innovation: Towards a Theory of Innovation and Interactive Learning*, London: Pinter, p. 6.
29. *Ibid.*
30. *Ibid.*
31. *Ibid.*
32. Shin, JS (1996) *The Economics of the Latecomers: Catching Up, Technology Transfer, and Institutions in Germany, Japan and South Korea*, London: Routledge, p. 33.
33. Patel, P & Pavitt, K (1994) 'National Innovation Systems: Why They Are Important and How They Might Be Measured and Compared', *Economics of Innovation and New Technology* Vol. 3, pp. 77–95.
34. *Ibid*:91.
35. *Ibid.*
36. Freeman, C (2002) 'Continental, national and sub-national innovation systems-complementarity and economic growth', *Research Policy*, Vol. 31, No. 2, p. 200. The source Freeman cites who developed the concept of passive and active learning systems is Viotti, VB (1997) *Passive and Active National Learning Systems*, PhD Dissertation; New School for Social Research.
37. Patel & Pavitt, 1994:77–95.
38. *Ibid.*
39. Freeman, 2002:200.
40. OECD (1992) *Globalisation of Industry: Overview and Sector Reports*, p. 20.
41. Ohmae, K (1990) *The Borderless World: Power and Strategy in the Global Market Place*, London: HarperCollins, xi. Also quoted in Freeman & Soete, 1997:307.
42. *Ibid*:199.
43. Freeman & Soete, 1997:307.
44. Porter, M (1990) *The Competitive Advantage of Nations*, London: Macmillan, p. 19. Also quoted in Freeman & Soete, 1997:307.
45. Saviotti, PP (1997) 'Innovation Systems and Evolutionary Systems' in C. Edquist (ed.) *Systems of Innovation: Technologies, Institutions and Organisations'*, London: Pinter, p. 196.
46. Scherer, FM (1992) *International High-Technology Competition*, Cambridge, Massachusetts: Harvard University Press.
47. Hirsch, P & Gillespie, JJ (2001) 'Unpacking Path Dependence: Differential Valuations Accorded History across Disciplines' in R Garud & P Karnoe (eds) *Path Dependence and Creation*, Mahwah, New Jersey: Lawrence Erlbaum Associates Publishers, pp. 69–90.
48. David, PA (1985) 'Clio and the Economics of QWERTY', *American Economic Review*, Vol. 75, No. 2, May, pp. 332–37; Arthur, WB (1989) 'Competing Technologies, Increasing Returns and Lock-in by Historical Events', *Economic Journal*, Vol. 99, pp. 116–131.

49. Ruttan, V (2001) 'Sources of Technical Change: Induced Innovation, Evolutionary Theory, and Path Dependence' in R Garud & P Karnoe (eds) *Path Dependence and Creation*, Mahwah, New Jersey: Lawrence Erlbaum Associates Publishers, p. 113.
50. David, P (1986) 'Understanding the Economics of QWERTY: The Necessity of History' in WN Parker (ed.) *Economic History and the Modern Economist*, Oxford: Blackwell, p. 30.
51. David, PA (1993) 'Path-dependence and predictability in dynamic systems with local network externalities: a paradigm for historical economics' in D Foray & C Freeman (eds) *Technology and The Wealth of Nations: The Dynamics of Constructed Advantage*, London: Pinter, p. 208.
52. Rosenberg, N (1994) *Exploring the Black Box: Technology, Economics and History*, Cambridge: Cambridge University Press, pp. 9–23.
53. *Ibid*:10.
54. *Ibid*:15.
55. *Ibid*:10.
56. *Ibid*:15–16.
57. *Ibid*:21.
58. *Ibid*.
59. Arthur, 1989:116–131.
60. This is the term used by Arthur.
61. Arthur, 1989:116.
62. *Ibid*.
63. *Ibid*.

3
Public Policy Frameworks: A Diverse Landscape

This chapter analyses the different public policy frameworks of Japan, South Korea, the US, and Western Europe.[1] There are two important aspects of these countries' policy frameworks which are relevant here. The first concerns their diversity. This is largely the product of the countries' different long-term priorities in the policy arena, and their very different histories. As we shall see in the following discussion, Japan's and South Korea's priorities have long been very different to those of Western Europe and the US. This leads us nicely on to the second point, which concerns the effectiveness of the different policy frameworks in terms of promoting technological innovation. Because of where they have focused their priorities, the policy frameworks of Japan and South Korea have proved far more successful at promoting technological innovation than those of Western Europe and the US. The implications have been profound. Primarily because of the enormous success which Japan enjoyed during the 1980s, Western Europe and the US changed certain elements of their own policy frameworks in the expectation that this would improve their record of technological innovation. At least in the field of electronics, these changes have not had the desired effects. When a country is faced with a major competitive challenge, changing elements of its policy framework is one of the usual measures adopted to address a problem of this nature. However, when the measures taken fail to work in the way that was originally envisaged, the country finds itself back at square one.

This is precisely the situation in which Western Europe and the US have found themselves in relation to TFT-LCDs. In terms of practical public policy measures they could take to enter the TFT-LCD industry, the room that they had for manoeuvre was very limited. At the same time, they face another major dilemma, what to do about South Korea

and Taiwan. In the present trade environment, where overt protectionism is now seen as being completely unacceptable, Western Europe and
the US now find themselves in the position where they have to think
of 'new' ways of tackling what is essentially an old problem, their own
lack of competitiveness.

With this in mind, the US, in particular, has attempted to move the
agenda on to what it perceives to be a much less apparent type of protectionism. The type of protectionism it has focused its attention on
in recent years is 'structural'. Various aspects of the industrial systems
of specific countries are believed to give them an unfair advantage
over the majority of other countries. As we have now come to expect
the main target for the US has been Japan, which it is convinced has
various trading practices that offer huge advantages to its domestic
firms. The US started to tackle this very thorny issue in 1989 when it
launched the Structural Impediment Initiative (SII). The original idea
was to reduce all general structural impediments affecting trade
between the two nations in all industries.[2] The reality of the situation
has been, however, that other issues have 'found' their way into the
negotiations, such as *keiretsu* relationships and the organisation of the
distribution system.[3] This formed part of a more concerted attempt by
the US to 'manage' its trade with Japan, as mentioned in Chapter 1.
The US continues to assume that if it pursues these sorts of public
policy measures for long enough, and can get Japan to come round
to its way of thinking, the country's competitiveness 'problem' will
gradually evaporate.

As yet there are no signs that the US is about to abandon this strategy. With the continuation of its huge trade deficits, and the corresponding large Japanese surpluses, the latest US plan to bring trade
more into line between the two countries is for Japan to implement a
comprehensive programme of liberalisation. The idea is that if domestic demand in Japan rises to a 'sufficient' level, it will boost US exports
and bring down its deficits. In spite of its domestic economy currently
being in the doldrums, Japan is not convinced this is the solution to
solving the US deficits or its large surpluses. The country has never
been enthusiastic about liberalisation, and there is no real reason for it
to be now. The US, on the other hand, has yet to realise it is still
missing the point.

As Tyson (1992) rightly points out, Japanese trade practices are not
the root cause of US trade deficits, and pursuing policy measures in the
form of the SCTAs and SII is no way to go about solving them.[4] What
the long and protracted tussle between the US and Japan graphically

illustrates is the ease with which home grown problems can lead a country to indulge in continual provocation of its trade partner(s), and allow its concentration to be completely diverted from focusing on the problems it really should be dealing with. From one standpoint, it is almost as if the US deliberately goes around trying to find the next 'unfair' trade issue to take up with Japan, or for that matter any other newly industrialising country (NIC) that is advancing rapidly, in order to rebuke them and giving grounds to the US to claim it is always the innocent victim of some conspiracy designed to cause it serious injury.

One reason why the US and Western Europe have to promote liberalisation in the way they do is because it forms a fundamental part of traditional economic thinking. As there is absolutely no way they can attempt to emulate the Japanese 'model', they are stuck with promoting an idea that is not always in their own best interests. Any moves which could be construed to be inhibiting the process of liberalisation are regarded as an anathema. With the tide of liberalisation continuing to flow strongly, pressures on Japan to move faster on the liberalisation of its economy will continue. Its reluctance to make any major concessions on the issue, however, is reinforced by the intense competitive pressures it is now under from South Korea and increasingly China. Japan is still pondering the question of whether it can be right to allow external pressure to determine whether it moves towards 'reforming' the national system of innovation (NSI) that has served it so well in the past? Clearly, the system needs to be refined to make it more flexible so it can cope with the growing pressures posed by globalisation. The trap Japan has to avoid at all costs if it is to maintain its competitiveness is not to damage its industrial base. Whether there is a strong link between liberalisation and a weak industrial base is hard to assess, but those economies that have proceeded the furthest down the liberalisation path have a weaker industrial base relative to Japan.

To say the least, the emergence of South Korea as a major competitive force in semiconductors has been bad news for Western Europe and the US. Now the country has managed a similar accomplishment in TFT-LCDs it threatens to present them with another headache that by all accounts could become quite acute. Hence, South Korea has found that it too has become an increasingly frequent target of US provocation of the type which has long been meted out to Japan. Faced with a maturing 'adolescent Japan', the US wants to avoid at almost any cost being on the receiving end of successive export drives. As liberalisation is one of the last remaining arrows the US still has in its bow to perforate South Korea's competitiveness, it is doing all it can to push this agenda. But like Japan before it, South Korea is not keen on liberalisation, and has been very slow to make various reforms that the US has been pressurising it to make, such as the reduction and elimination of certain tariffs. It has proved just as reluctant as Japan to let external pressure determine what reforms it should or should not make. There is no logical reason for the country to be pushed into making changes to its NSI, which has taken the country to the technological frontier in two key sectors.

Now that Japan has become accustomed to US trade antics, and South Korea has had to learn fast about them, the respective conclusions of the two countries must be that the US has an intractable problem which it cannot solve. While the US has been grabbing the limelight, the EU has been happy to take a back seat, content to see how the US has got on, and then decide what action it may need to take. From its vantage point, the failure of US efforts to stifle Japanese and South Korean competitiveness through various policy initiatives will have important implications for its future strategy. The EU is certain to conclude that if the US cannot succeed, there is no point trying any similar measures of its own as they are likely to be no more successful.

The only area where Western Europe and the US have felt comfortable enough to 'copy' Japan, if that is what it can be called, are changes they have made to some of their policies. A comparative analysis of Japanese, South Korean, European and US policies relating to competition, trade and technology is therefore useful. These policies can either have a positive or negative impact on the long-term survival and prosperity of individual sectors. Ascertaining which policy framework has been the most effective is necessary as the tensions between the objectives of these policies and their ongoing implementation have been on the increase in recent years, due to globalisation and the increasing involvement of governments to promote innovation and technological progress (Mowery & Rosenberg, 1989, Jorde & Teece, 1992, Holmes, 1993, Buigues *et al.*, 1995).[5]

Policies relating to competition, trade, and technology

It is fair to say that all of the advanced industrialised countries (AICs) have policies in one form or another which relate to competition, trade and technology. They, of course, pursue their respective policies with one main objective in mind, their national self-interest. However, what is not immediately apparent are the huge differences that exist between countries in the way that they implement their respective policies. It is worth noting that these differences are observable in all three areas. If we take the area of competition first, it is viewed from very different perspectives, for example, by the US and Japan. As we shall see below, the US equates competition with free and open markets, where firms compete with each other on a level playing field and where the consumer is king. The US approach to competition is ideological, based principally on the *laissez-faire* ideas of Adam Smith. The main method which the US uses to maintain free and open markets at the national level is through its system of antitrust, which dates back to the late nineteenth century. Compared to the US, Japan's approach to competition could not be more different. From one standpoint, it could be argued that Japan's approach is a more practical one. Unlike the US, which has preferred to focus on promoting competition in markets, Japan has tended to focus almost exclusively on promoting competition between producers. The government reasoning behind this is quite straightforward. If the country's firms were to stand any chance of competing successfully at the global level, they first of all had to learn how to compete at the national level. If a firm could not compete at the national level where competition is usually (though not all the time) less intense, how could it be expected to compete at the global level? In essence, firms have been encouraged to learn how to walk before learning to run.

In so far as Western Europe and South Korea are concerned, Western Europe has a broadly similar approach to competition to that of the US, and South Korea has an approach not that dissimilar to Japan's. However, despite these apparent similarities it should not be assumed in any way that Western Europe attempts to promote free and open markets with the same vigour as the US does, or for that matter enforces its competition policy in the same fashion as the US enforces its antitrust policy. Likewise, South Korea has not tried to promote competition between producers in the same way as Japan has done.

In the field of trade, as in the field of 'competition', the approaches of Western Europe, the US, Japan and South Korea exhibit enormous diversity. The diversity of their approaches is largely a by-product of each country protecting itself from foreign competition in areas where they each feel themselves to be particularly vulnerable. It is important to note, for example, that the areas where the US thinks it is vulnerable are not the same as those of Western Europe. The same is equally true of Japan and South Korea. The two countries seek to protect themselves in very different (and unrelated) sectors.

Of all the countries, Japan has perhaps the least active trade policy. Because of its huge success across a wide range of industrial sectors, it has experienced little (or no) need to protect itself from foreign competition either because it has not proved that troublesome or it simply does not exist. That said, it should not be taken to mean that Japan has no need for a trade policy. In the section on Japan, it is stated that like any other country, Japan is willing to protect its national interest as and when the need arises. While Japan is relatively inactive on the trade policy front, the US is the most active country in this field. Unlike Western Europe, Japan, and South Korea which by and large use trade policy to protect themselves in specific sectors, the US has a very distinct approach. As discussed later on, the US regularly uses trade policy to attempt to prise open the markets of some its major trading partners, which it deems to have closed markets, for the benefit of US firms. One other important aspect of US trade policy, which is used in conjunction with this strategy of trying to open closed markets, is the extensive use of anti-dumping measures. Whenever US firms find that they are struggling against foreign competition in a sector, they have a tendency to file an anti-dumping petition against their foreign competitors, citing 'unfair competition' as the reason. US firms seem to think that if they file an anti-dumping petition it will somehow blunt the competitiveness of the foreign firms, but US firms have yet to appreciate that anti-dumping petitions do not resolve the problem of their own lack of competitiveness, and can create far more problems than are normally anticipated.

In contrast to the US, Western Europe and South Korea pursue their trade policies along more conventional lines. There are two main areas where the EU uses its trade policy, in mature industrial sectors, where Western Europe thinks European firms need some breathing space to restructure, and also in high-technology sectors, where it is weak. South Korea, on the other hand, has an even more selective approach to trade policy than the EU does. At one end of the spectrum, South Korea uses it to completely shield an expanding domestic industry from foreign competition. On the other, trade policy is conspicuous by its absence when the country needs to obtain vital industrial equipment from overseas which is necessary for maintaining the vitality of the country's industries. From the above, it can be seen that trade policy comes in several different forms. It is all too easy to make the assumption that trade policy is a blunt instrument widely used by countries to protect themselves, with a high degree of uniformity between their approaches. The reality of the situation is far removed from this. In the following sections on Japan, South Korea, Western Europe and the US, the areas where they apply their trade policies are specified and the reasons why.

Lastly, as we might have come to expect, the technology policies of Japan, South Korea, Western Europe and the US exhibit exactly the same major characteristic as their trade policies do; there is enormous diversity between their respective approaches. Once again, the diverse approaches can be largely attributed to the countries pursuing their own national interests. One key difference between a country's trade policy and its technology policy is that it really matters when a country started to pursue its technology policy. It is a country's technology policy rather than its trade policy that can significantly affect its economic performance. What is important here is that both Japan and South Korea started to pursue their technology policies much earlier on than either Western Europe or the US, and this has had a direct impact on their competitiveness. One has to look no further than the electronics industry for evidence of this.

It will be shown that technology policy has been of far greater importance to Japan and South Korea, than either competition and trade policies. Whilst these policies are obviously important and have their rightful place, the governments of Japan and South Korea appreciated much sooner than Western governments how valuable a well structured and focused technology policy is when it comes to trying to reach the technological frontier. For the US, on the other hand, competition and trade policies have long been at the core of the government's strategy to stimulate dynamism in technology-intensive industries, and to protect them. It was not until the early 1990s that the country finally started to move towards developing a technology policy, but the process has proved to be very slow due to the nature of the necessary reforms, and the strong anti-interventionist and antitrust traditions of the country. For Western Europe, all three policies are regarded as being vital to its economic well-being, but they have long been coordinated in a fragmented fashion, so they have consequently not been particularly successful. Western Europe and the US have been keen to develop a coherent technology policy, as it is a far more effective way than combining trade and competition policies to overcome foreign competition. Technology policy can be an effective mechanism for overcoming foreign competition as it focuses on developing and improving a country's NSI. As Japan and South Korea have amply demonstrated to Western Europe and the US, when a country's NSI works efficiently it can bring substantial rewards.

Constructing a public policy framework with just the right equilib-
rium necessary to reach the technological frontier in a specific industry
is very hard to achieve. Overall, Japan and South Korea have managed
to achieve a better balance than Western Europe and the US in this
area. By focusing on technology policy, Japan and South Korea have
had much greater control over the outcome of their respective policies
as they have been formulated and implemented at a national level. The
huge advantage for government at this level is they have influence
over various aspects of policy (education and training, promotion of
generic technologies and R&D), which thus makes it much easier to
sort out any problems that will inevitably arise.

Trade and competition policies though are subject to external variables that are usually outside national control. In the case of trade
policy, it is formulated at a national level, but it is implemented at an
international level. In Chapter 9, it is mentioned that when the US
filed an anti-dumping petition against Japan in 1991 as part of its
'strategy' to enter the TFT-LCD industry, the US discovered to its cost
that trade policy has a 'boomerang' effect. The theory is that if it is
properly implemented, it will hit a country's competitor(s) in the area
at which it is aimed, but the great tendency is for it to come back and
affect the country applying it in a negative way. The application of
competition policy, on the other hand, has been turned on its head at
both the national and international level by technological change.
This is borne out by a paper by the Organisation for Economic
Cooperation and Development (OECD): there is widespread recognition that the application of competition policy has become a very
grey area.[6] It highlights the challenge facing economists and countries
alike, the development of an analytical framework that neither
impedes desirable innovation nor permits market power to reduce
technological progress.[7]

In an attempt to bring greater clarity to the situation, one of the
main ideas put forward for future debate is that instead of applying
competition policy at a sectoral level, it could be applied to different
kinds of product markets. The three types of markets suggested are
current generation products, innovation markets and future generation
products.[8] By focusing on these types of markets, it is hoped it will
become easier to make proper and perhaps fairer assessments of very
complex cases. The biggest drawback of this approach that has yet to
be resolved is how to define 'the market'.[9] As innovation continuously
changes the composition of products, the challenge ahead is to find a
suitable definition.

Japan[10]

Competition policy

A common allegation made against Japan is that it has taken a lax approach to the enforcement of competition policy (Okimoto, 1989, Tyson, 1992). If a comparison were made with the US this would be true. Behind this allegation though lies a huge cultural and ideological gulf between the two countries over where the benefits of competition should be directed. In the US, where the commitment to the free market is unquestioned, the principle of consumer sovereignty is solidly entrenched.[11] In contrast, consumer interests have never been placed on such a high pedestal in Japan. In line with what is best for the public interest, the interests of producers are given priority. In order words, the theory is, 'What is good for Japanese business is good for Japan as a whole and therefore good for individual consumers'.[12] This has very similar overtones to the old phrase, 'What is good for General Motors is good for America'. The difference is, though, the interests of producers have been enshrined in legislation. In view of the origins of competition policy and the circumstances under which the system was established this is not a total surprise.

The basis of the country's competition policy is the Anti-Monopoly Law (AML) enacted in 1947, and administered by the Fair Trade Commission (FTC). It was a historic piece of legislation as, prior to 1945, no legislation of this kind existed. The AML was modelled on strict US antitrust laws, mirroring US policy towards Japan at the time, which was to make it a strong and democratic country. Enforcement was a major priority before Japan regained its independence, but once it had control over its own affairs the AML was made more applicable to meeting the needs of the devastated post-war economy.

In 1953 a series of amendments were made to the AML so it would not become a hindrance to reconstruction. One of the most far-reaching changes in policy was in relation to Article 4. The outright ban on cartels *per se* was abolished and they were allowed on the grounds that they did not cause a substantial restraint of competition in a particular field of trade.[13] Furthermore, 'rationalisation' cartels and 'depression' cartels were granted legal exemption from the AML with authorisation from the FTC. Both types of cartel were ultimately aimed at maintaining stability in an industrial sector when the business cycle either turns in its favour or against it. Another key amendment was made in respect to acquisitions and mergers. Again, these were permitted so long as they did not substantially restrain competition in a particular field of trade. These amendments sufficiently weakened the AML to allow the government the vital room for manoeuvre it required for the coordination of important policies, such as industrial policy. The AML was not subject to any further reform until 1977 when measures were adapted to strengthen its enforcement, especially in the area of monopolisation.

Trade policy

As far as trade policy is concerned, the literature gives the impression that Japan has little need for it as most of the trade disputes it has with Western Europe and the US are over high-technology industries. In almost every instance Japan is portrayed as the culprit. In areas where Japan is vulnerable though, it is as prone to using trade policy as any other nation, for example, in the protection of its domestic rice market from foreign competition.

Technology policy

From a very early stage, Japan recognised the enormous benefits of pursuing a well-defined technology policy well before any other country did, and that it only functions effectively under specific conditions. Arguably, the first element of the policy was the passage of the Electronics Industry Promotions Law in 1957, which gave the government, amongst other important powers, the necessary freedom to encourage interfirm collaboration to develop key technologies free from the fear of potential prosecution. The other elements of the policy comprised of developing a highly skilled workforce, the import and internal diffusion of foreign technology, and the creation of appropriate policies which encouraged firms to take up radical innovations which had a long gestation period before they became profitable. The main strengths of the policy are the length of time it has been in operation and the range of measures in key spheres which are designed to encourage technological innovation.

South Korea

Competition policy

The government did not consider the issue of competition policy an important one until 1980. It is discussed in Chapter 8 that prior to that date the government was heavily preoccupied with the country's economic development. It rapidly changed its mind on the issue in the wake of the growing economic power and influence of the chaebols to which it had given enormous assistance to develop and expand. As might be expected with any group of businesses that have become very large, the government discovered the chaebols were guilty, amongst other things, of monopolistic abuses, predatory behaviour, and price gouging.[14] In an effort to combat these growing problems and rein back the chaebols' economic power, the government passed the Monopoly Regulation and Fair Trade Act (MRFTA) (1980) which is modelled along the lines of US antitrust laws.[15] In theory, the Act prohibits unfair cartel practices and mutual investment among the chaebols' affiliated companies; places restrictions on their vertical and horizontal integration; and lastly, places a ceiling on investment by and credit to the large chaebols. In addition, the government directed the 30 largest chaebols to restructure their businesses and concentrate on three or fewer core areas.[16]

Eager though the government might have been to curb the excesses of the chaebols, and to make them act more responsibly, time has proved the MRFTA is nothing more, to borrow a phrase of Chairman Mao's, than a 'paper tiger'. The deep and intimate links that have long existed between government and industry meant that during the 1980s, the chaebols could carry on with their habit of growing, irrespective of the problems it was causing in the wider economy. Kim (1997) provides evidence that shows beyond doubt the extent to which the chaebols had the upper hand over the government. For example, out of 1172 applications for vertical and horizontal integration the government only rejected two.[17]

Against the backdrop of the continued growth of the chaebols, and the changing global economic environment, in December 1992 the government revised the MRFTA. The reform centred on three issues, the introduction of limitations on cross-debt guarantees between affiliate companies, exemptions to limitations on the amount of total investment in other companies by the chaebols, and lastly, enforcement provisions.[18] Kim contends this was not an attempt by the government to give itself the necessary powers with which it could stringently regulate the activities of the chaebols, but was a move in the opposite direction. So that the chaebols could successfully compete at a global level, the government made a marked policy shift away from regulation to liberalisation. Nowhere was the change in emphasis more visible than in the removal of credit controls on the chaebols. Provided they complied with specific criteria, they were effectively left free to pursue their ambitious investment programmes, which many of them subsequently did.[19]

With the government having been in cahoots with industry for so many years, in reality the government has never had an opportunity of pursuing an effective competition policy without fear of upsetting the apple cart, which it has proved extremely reluctant to do. On the basis of its past record, the only way the country will be able to have an enforceable competition policy that has any substantial bite will be through a mixture of external and internal pressures which are thrust upon it. The most likely source of external pressures are going to be the OECD to which South Korea now belongs and its key trading partners, especially the US and EU. The internal pressures are going to come from the need for more balanced economic development, most notably the development and growth of small and medium-sized firms. This has been put off up till now, but eventually the issue will have to be addressed whether the country likes it or not.

Concern over the nature of the MRFTA reforms on domestic competition policy and the economy can be found in a WTO report published in 1996. It states, 'While the reforms are conducive to long-term economic resilience...there may be no immediate gain as large industrial conglomerates are well represented in dynamic and innovative industries. Trading companies related to such conglomerates handled over 46 per cent of Korea's exports in 1995, up from 38 per cent in 1990'.[20] What this implies is that if nothing else, any attempts made by the government to reverse deeply entrenched, long-term trends will prove a struggle.

Trade policy

Until the 1990s the country's trade policy was a highly protectionist one. With South Korea's desire to fully integrate itself into the global economy, and having become a member of the WTO and OECD, its policy of protectionism has started to seriously fray at the edges. In an ideal world it would no doubt liked to have to continued with it, but under pressure from these bodies, it has had to agree to start moving towards much greater liberalisation of its economy. As discussed earlier, liberalisation is not a process the country either finds very attractive or easy to come to terms with. South Korea might be under external pressure from the WTO, the OECD and the US, but it is only proceeding with liberalisation at a rate which it considers appropriate, that is, slowly.

The area where liberalisation has progressed the furthest has been in the manufacturing sector. In order to conform with the multilateral trading rules laid down by the WTO, tariffs have been reduced across virtually all sectors. With its Tariff Reduction Plan, the average tariff has been reduced from 18.1 per cent in 1988 to 7.9 per cent in 1994.

The average tariff on manufactured goods was 6.2 per cent on a trade-weighted average in 1995. These external pressures have also led to restrictions on certain Japanese products, maintained under the Import Diversification Programme, being gradually phased out and this was completed ahead of schedule, prior to the 1999 deadline.

At the sectoral level, the government continues to use trade policy on a 'selective' basis. In the car industry, which has grown steadily in recent years, trade policy has been used to give it almost wholesale protection from foreign competition. Imports of foreign cars have been kept at a very low level.[21] Trade policy's more 'liberal' side, so to speak, can be seen in the electronics industry. In its rush to reach the technological frontier, the government has not been able to use the same approach as it has used for cars. Vast amounts of capital goods and electronic components have had to be imported, especially from Japan, as Korean firms needed to get their hands on the most modern equipment they could. As painful as it may have been, internal pressures forced the government into taking this course of action. The truth of the matter was that the country simply did not produce the capital equipment and components it required, so was forced to import them. In the future, there will be piecemeal liberalisation of the economy, but South Korea will do it at a pace that suits it and avoids damaging its industrial base.

Technology policy

The most enduring feature of South Korea's technology policy has been the way in which the government has maintained a very tight grip over both its direction and focus. In true Korean tradition, the government has followed the example set by Japan, of focusing on both the supply and demand sides, so as to make its policy as effective as possible. On the supply side, the government paid particular attention to policies which helped strengthen the country's scientific and technological capabilities. One of the central planks of the strategy to get the process under way was initially to rely on foreign technology imports from abroad, mainly from the US and Japan, to lay the foundations for its own indigenous industries.

The government realised that whilst this policy had much going for it, it suffered from two serious deficiencies which meant that the country could not rely on it indefinitely. Firstly, too great a reliance on foreign technology imports could stunt the development and growth of a domestic industry, as in the case of the capital goods sector, which saw a massive growth in imports from the US, Japan, and elsewhere between 1962–93. Secondly, the reliance on technology imports was not of much practical assistance if the country was to be able to move into technologically advanced industries of its own accord and successfully compete in them. As the chaebols discovered in the semi-conductor industry, there are limitations to the assistance that potential competitors are prepared to provide to help any one firm establish itself in a sector.

To achieve the latter point, it was decided the country had to build up its own R&D capability. The government decided to tackle this immense challenge with the creation of a number of government research institutes (GRIs) such as the Korean Institute of Science and Technology (KIST), which was established in 1966.[22] In an effort to achieve its industrial objectives, the government established a number of further GRIs, which were spin-offs from KIST. These were designed to build up capabilities in specific sectors which had been given priority by the government. These GRIs were supplemented by the creation of two science parks, the Seoul Science Park, established in the same year as KIST, which had three R&D institutes and three economic research institutes, and the Taedok Science Park which was created several years later in 1978, which now has 14 GRIs.[23] It is worth mentioning here that about 90 per cent of government research contracts for its key projects such as its National R&D Project (NRP) and Industrial Generic Technology Development Project (IGTDP) are awarded to the GRIs.[24]

The proliferation of the GRIs has been intended to compensate for the country's weakness in basic scientific research. This is at least one area of its technology policy where the government has not managed to 'get it right'. Although the problem is being sorted out, the country's universities are overwhelmingly oriented towards undergraduate teaching, and with that has come a distinct lack of a research culture.[25] Changing the culture and quality of the universities was always going to be a difficult task, and it has proved just that. In anticipation of this, in 1975 the government set up the Korean Institute of Science which offers research-oriented Master's and PhD programmes. In 1995 this was supplemented by the creation of another such institution. These two institutions now account for approximately half of all science and engineering PhDs in the country.[26]

On the demand side, the government has focused its attention almost exclusively on the control of the chaebols. It is discussed in Chapter 8 that the government has encouraged the chaebols to enter sectors where the risks are high. The underlying objective has been to make sure that technological innovations are successfully exploited. The chaebols have been only too happy to oblige because if they were successful there was not only the prospect of jam today and tomorrow, but well into the future! Whatever criticisms there might be of this policy, it has had the desired effect of propelling the country to the forefront of semiconductor and LCD technologies.

European Union

Competition policy

When the Treaty of Rome was ratified by the six original member states to launch what was then known as the European Economic Community (EEC) in 1957, it was clear that competition was to have a very important role in its future. For the EU to thrive over the coming years, it was deemed necessary that it ought to have its own system of free and fair competition. To guarantee the creation of a system, a specific provision (Article (3)) was inserted into the treaty. Whatever the theoretical niceties the architects of the original treaty might have had about free competition between the member states, the reality of the situation, as with many other things in this world, was that the issue of competition remained basically dormant until the early 1970s. In the intervening period, the member states had had their attention focused elsewhere, predominantly on the restructuring of European industry. What brought the issue of competition to life in 1972, was a proposed acquisition of a European firm by a US firm, Continental Can, whose objective was to establish a dominant position in the European canning industry, which the EU was anxious to block.[27]

The Continental Can case acted as a catalyst in two ways. Firstly, it made the EU sit up and take notice of the need for a coherent policy, as there was no effective mechanism to speak of that could monitor how the competitive environment varied between the member states. And secondly, it exposed one of the inherent weaknesses in the existing system, the lack of proper merger controls. This helped knock on the head the idea that creating a system of free competition was going to be straightforward. Since the early 1970s, strenuous efforts have been made to make competition within the EU as free as possible, but the problems which surfaced in the early 1970s have not gone away. Although there is a Directorate (DGIV) in the Commission that has the responsibility and appropriate powers to regulate competition throughout the EU, the competitive environment between the member states remains far from 'harmonised'. As Gilchrist and Deacon (1990) show,

competition has been distorted for many years by member states channelling large subsidies (or state aid) into predominantly major public-
sector firms, and into 'sensitive' industrial sectors like steel,
shipbuilding, coal, textiles, and cars, experiencing difficulties.[28] There
are provisions in the Treaty of Rome, Article 92 and Article 93, which
gives the EU the necessary powers to regulate the flow of subsidies. In
recent years, the EU has sought to turn the flow of subsidies into a
trickle because of the way they can and do distort competition, but it
has only had limited success.

A combination of huge political pressures and several important
exemptions to the provisions has made the task of holding back the
flow of subsidies an unenviable one. Article 92 bans 'any aid granted
by a Member State or through State resources in any form whatsoever
which distorts or threatens to distort competition by favouring certain
undertakings or the production of certain goods'.[29] This appears to
make the granting of subsidies a stringent process, but the reality of
the situation is quite different, as the list of exemptions has given
the EU and member states a considerable degree of leeway. Some of
the exemptions include: aid which promotes the economic development of areas suffering from abnormally low standards of living or
serious underemployment; aid which promotes important projects of
a common European interest, or remedies a serious disturbance in the
economy of a member state; and finally, that which facilitates economic development in ways which do not conflict with the common
European interest.[30] With these and other exemptions, the principle
of competition has been compromised on many occasions as other
objectives of the EU have been deemed to be more important.

Whereas subsidies to the 'old' industrial sectors have been relatively
easy for the Commission to keep a close eye on, subsidies for research
and development, known as 'horizontal aids', have proved a much
harder problem to solve.[31] The mechanisms used to distribute them
are not very transparent, and to make the Commission's problems
even worse, government intervention in this area has increased
because they can be 'wrapped up' for the promotion of the 'technical
or economic progress' of Western Europe, one of the major goals of
the EU. This seemingly innocuous phrase, which is stated in Article 85
(3), has proved to be of major significance. In recognition of Western
Europe's precarious technological position, the EU has long been
aware of the need to continuously seek ways to improve it, and has
hence championed the cause of high technology industries. As far
back as 1972, the EU recognised its importance, and, in reference to

specialised agreements acknowledged, 'Such agreements do...provide a means of obtaining a specialisation which contributes to lower costs by the setting up of long production runs and a better utilisation of available capacity by the concentration of effort on a limited number of products'.[32]

Concern over the continued use of subsidies to promote the economic or technological progress of the EU is very real. Sharp & Pavitt (1993) point out that unless the EU has a strong competition policy, where several large firms (without the support of subsidies) compete vigorously against each other, then there is a real possibility that the mistakes of the national champion strategies will be repeated at the European level.[33] Their fears seem to be very well grounded. It has been alleged that what was Europe's only LCD producer, the Flat Panel Display Company (discussed in more detail in Chapter 9), not only received subsidies from the EU to help set it up, but also received additional funds to help keep it afloat because of the severe difficulties experienced. For both economic and political reasons, the EU felt it could not afford to let an important project of this nature fail after such a short period of time.[34] Subsidies are something that the EU has been grappling with for years. The idea of the EU has been to steadfastly refuse to go down the same road with the 'sunrise' industries as it did with the old 'smokestack' industries. From hard experience, all the EU found was that it created a mountain of problems of its very own which were in the end in no party's long-term interest.

Where competition policy is on a far more solid footing is in the extension of exemptions to high-technology industries, which has brought it much more into line with Japanese competition policy. The process of giving greater exemption to high-technology industries started in the early 1980s, as it was realised it made no sense trying to regulate such industries at a national or European level if they have to compete globally in order to survive long term. Collaboration for the purposes of 'pre-competitive' research is granted a block exemption under Articles 85 and 86 of the Treaty of Rome, but competitive research at the development stage does not have one.[35] Recognising that this might be a barrier to further technological innovation, in 1984 the EU introduced its block exemption for certain joint research and development agreements. This was later updated by the 1993 block exemption of such agreements for joint ventures to market the products resulting from the joint research and development.[36]

The value of these exemptions is illustrated by the Flat Panel Display Company. Without the exemptions in place, it is highly doubtful whether Philips and its partners would have bothered with the venture, if they knew that in addition to the high costs they would incur setting it up, they would also be potentially liable to prosecution which could leave them being saddled with huge fines. In the event, the EU was able to approve the venture on the grounds that, 'the joint venture is going to be a means for the parent companies to develop and sustain mass-production in Europe of new high-technology products in a field where competition from foreign suppliers (mainly Japan) is strong',[37] but the question of subsidies threatens to offset any potential gains. Perhaps aware of this acute dilemma, the EU actively uses trade policy to strengthen those sectors in which it is weak.

Trade policy

Trade policy is used widely by the EU when it is faced by two common, but very different set of circumstances. It has long been used to help alleviate competitive pressures on European firms in declining and mature sectors that have been badly affected by foreign competition. In the car industry for example, the EU has long played an influential role in protecting the indigenous industries of the member states from the worst effects of Japanese competition by negotiating on their behalf with Japan and devising ingenious protectionist measures. From the mid-1970s, when Japanese cars started to flood on to the European market, the EU regulated the flow of imports from Japan until 2000, primarily through Voluntary Export Restraints (VERs) and 'consensus agreements'. These were designed to give the European firms the necessary time to restructure so they could reach Japanese levels of efficiency and productivity.[38]

Trade policy takes on a different dimension when it comes to an industry like semiconductors. In Chapter 4, it is mentioned that during the 1960s the UK and French governments had a policy of encouraging overseas investment, particularly by US firms. This source of investment has since been complemented by Japanese and South Korean investment in the sector. Here trade policy has been used to strengthen a sector it would otherwise be weak in. The truth of the matter is, even with its size and diversity the EU does not have a sufficient number of firms that can compete successfully at a global level.

Technology policy

Moves to develop a European technology policy gathered speed at the beginning of the 1980s as a result of two major influences on the EU. The first, and most potent, was the success of Japan's 1975 VLSI (very

large scale integration) semiconductor initiative, and the announcement of its successor in 1982, the Fifth Generation Computer Programme.[39] The second was Vicomte Davignon, Commissioner for
Industry (1977–85), and for Research and Technology (1981–85), who
was well aware of the possibility of Western Europe falling behind the
US and Japan in the development and exploitation of new technologies.[40] Against the backdrop of Japan's success, and Western Europe's
more general competitiveness problem, Davignon made it almost his
own personal mission to create the foundations for a European-wide
collaborative research system. To get the process underway, Davignon
used the VLSI example as his model, and then went about getting the
heads of Europe's leading electronics and IT firms to engage in a series
of discussions to find out their priorities and to ascertain how committed they were to the idea of future programmes. By doing things this
way, Davignon was able to prevent his initiative from getting tied up
in red tape and failing. At the time, the normal route for initiatives of
this nature was to find their way as high as the Commission's research
directors, only then to flounder because they did not move further up
the hierarchy where the power to take them forward lay.[41]

As a result of Davignon's hard work, an outline programme of pre-
competitive research had emerged by September 1980.[42] The idea was
to develop a programme based on collaboration among the major
European companies, small and medium sized firms, research institutes
and universities.[43] Little more than a year and a half later, a complete
proposal and the Commission's paper, 'Towards a European Strategic
Programme for Research and Development in Information Technology'
(ESPRIT), were ready for approval by the Council of Ministers, and later
at the European Summit in June 1982.[44] Once they had been approved,
the Commission was then in the position to implement them. Within
six months of approval being given, the first pilot projects had been
given the green light at a cost of £8.5 million.[45] This gave the Commission the opportunity to see how the process of collaboration actually worked in practice, which was vitally important, given that large
amounts of European taxpayers' money was at stake.

A total of 38 projects were launched during the pilot phase (1983–84),
which at the end of the period were considered to have been a great
success. The success of the pilot phase led the Commission to
go ahead with its plan for a full ten year programme (1984–93), which
was financed with 50 per cent of the funding coming from the
Commission and the rest from industry. The total amount of funding
from both of these sources reached ECU1.5 billion, approximately

£1 billion. The first phase of the programme, which spanned a period of five years, concentrated on microelectronics, advanced information processing, software technology, office systems and computer integrated manufacturing.[46] The response, which the Commission had from its call for research projects, was so good that ESPRIT got off to a flying start, signalling that an EU technology policy had finally been born. The success of ESPRIT led to intense pressure for more comprehensive and expensive programmes in the future, which the EU has struggled to keep within reasonable limits. Nevertheless, a widespread belief took hold that the EU had at last found the right formula for its technology policy, which meant that it was finally given legal sanction in the Single European Act, passed in 1987. In essence, it is an update of Article 85 (3), as Article 130F of the 1987 Act states, 'The Community's aim shall be to strengthen the scientific and technological base of European industry and to encourage it to become more competitive at an international level'.[47]

More than a decade after the birth of the technology policy, questions have been raised about its effectiveness. The original belief that a comprehensive research programme, the cornerstone of the EU's policy, would give a much needed boost to European industry and lead to some kind of an industrial renaissance has given way to mounting scepticism. The main criticisms levelled at programmes like ESPRIT which fall under the rubric of the Framework Programme, is that they have not measurably improved Western Europe's trade deficit in electronics with Japan, and nor have they led any large European firms to become major global players. Whilst these criticisms are very valid, the main counter argument is that the EU cannot afford not to develop promising new technologies, just for the sake of saving a small amount of European taxpayers' money. So far the latter argument has not been seriously challenged and the outlook for the EU's technology policy looks quite good. Its prospects though might not be so good in future years if it does not start yielding the results it was set up to produce in the first place.

The delay in developing a technology policy was caused by two major factors, one of which has taken a long time to remedy, and the other, which for all intents and purposes, never will be. The first concerned the lack of a policy framework in the Treaty of Rome, which from the very creation of the EU denied the Commission the powers to promote research and development, with concomitant repercussions.[48] The second factor, which is of an even more fundamental nature, concerns the deep ideological divisions between the member states who were determined to pursue their own national interests. These ideological differences, which existed between the member states in the 1960s and 1970s and are still very much in evidence today, focused on the merits of free competition versus interventionism. The clash of cultures centred around the strong German sense of liberalism and the equally strong French sense of interventionism.

Interventionism was for many years the EU's dominant philosophy. What made interventionism so appealing was that when US corporations dominated the European corporate landscape throughout the 1960s and early 1970s, European governments equated industrial success with size of the firm. We now know of course that this was a fundamental mistake. The national champions' strategy of making large firms even bigger so they could compete with predominantly US firms, was by most standards a disaster. Not until the situation had reached crisis point, however, did both the philosophy and focus of support change. Despite the national champions' strategy having fallen from grace, liberalisation is still not as revered as interventionism was, least of all by France, which still remains deeply committed to the interventionist principle. This continuing French enthusiasm for interventionism has frustrated the efforts of the EU to speed up the process of liberalisation of certain sectors, for example, energy. Liberalisation might now be the dominant philosophy, but the legacy of interventionism still runs deep as the very slow pace of reform testifies.

United States

Competition policy

As is well known, the US competition policy or antitrust system rests on the Sherman Act (1890) and the Clayton Act (1914). From around the turn of the twentieth century through to the 1970s, the contribution of the antitrust system to the country's economic welfare was never seriously challenged. This led Neale (1970) to state:

> The antitrust laws of the United States of America are unique in their scope of content and rigour of enforcement. In no other country is there an elaborate and comprehensive body of law dealing with monopolies and restrictive practices. At the same time it is widely conceded that the United States is pre-eminent in the power of her industry and commerce. It is natural, therefore, to infer some connexion between economic success and the existence of this special body of law.[49]

After the mid-1970s, serious questions were asked about the efficacy of the antitrust system as the country's trade performance started to falter, and it saw its technological leadership being challenged by Japan. It would be wrong to presume the antitrust system worked without any serious hitches prior to the mid-1970s. In a very good critique, *The Antitrust Paradox: A Policy at War with Itself*, Bork (1978) argues that prior to this date fundamental flaws in the system were already very much in evidence.[50] According to Bork, the flaws emerged as a consequence of the very narrow criterion used to evaluate competition cases and the way the criterion was interpreted by the courts. This led to at least two judgements concerning cases that made no logical sense. By themselves these judgements did not lead to any radical reform of the antitrust system as a whole, as it was otherwise assumed to be working well. Calls for reform were only given consideration once it had become apparent that the country's economic 'invincibility' was under severe attack and the words of Neale (1970) began to ring rather hollow.

The US is like no other industrialised country in the sense that it has a long and enduring tradition of popular hostility towards big business generally.[51] The foundations of this of course were laid by the infamous 'Robber Barons', who dominated corporate America from the latter part of the nineteenth century to around the 1920s. Their actions led to the passing of the Sherman Clayton acts, to prevent firms which were intent on trying to obtain a monopoly position in an industry, or from abusing their monopoly if they already had one. With the passage of the two acts, the prime objective of antitrust became the maximisation of consumer welfare,[52] as it quickly became the conventional wisdom that big business was always out to 'rip off' the consumer.

Indications that the excessive emphasis of antitrust policy on consumer welfare was a disaster waiting to happen came as early as 1945 with the Alcoa (*United States v Aluminium Co of America*) case. Alcoa was a firm that over a long period had steadily built up a dominant position in 'virgin' aluminium ingots. Its dominant position originated from a patent monopoly, which it had enjoyed from 1899–1909.[53] Over the following 28 years the firm successfully managed to keep any potential competitors at bay, and by 1940 had a market share estimated to have been approximately 90 per cent.[54] Its market share was a consequence of its productive efficiency and nothing else. After its patents had expired it had no protection from other firms that wished to enter the industry, so the only way it could maintain its position was to be highly competitive. Bork claims the architects of the Sherman Act were not against firms that obtained a monopoly through efficiency, but those obtained through either horizontal mergers or predatory behaviour.[55] In 1945, in the Court of Appeals, Alcoa was judged to have violated Section 2 of the Sherman Act as its share exceeded more than two-thirds of the market. The judgement effectively meant that growth through entirely appropriate means was illegal, and to make matters worse the court never identified who the 'customer' was it sought to protect from Alcoa.

The Alcoa judgement was followed by another one that made a complete mockery of the country's merger policy. In 1962 there was a case, *Brown Shoe Co. v United States*, before the Supreme Court, which under normal standards would not merit a second look. It concerned two firms in a highly competitive industry that wanted to merge. Brown, a firm that primarily manufactured footwear wanted to acquire G.R. Kinney & Co, which was primarily a footwear retailer. Their respective shares of the nation's shoe output were 4 per cent and 0.5 per cent. Kinney had 1.2 per cent of total national shoe retail sales by dollar volume (no figure was given for Brown), and together they had 2.3 per cent of total retail shoe outlets.[56] One might have expected with these very small market shares in an industry with more than 800 manufacturers that the merger would have been given the green light. The Court, however, somehow found the acquisition was a threat to competition at both the manufacturing and retailing levels and declared it illegal.[57] As Bork points out, this type of judgement did an immense amount of damage to the domestic economy by inhibiting or destroying a broad spectrum of useful business structures and practices.[58] There was no real incentive for corporate America to radically change the way it functioned as long as it had the very dark cloud of litigation constantly hanging over it.

When competition from Japan started to hit the US hard, pressure mounted on the country to reform its antitrust system, which had gone unchanged for more than sixty years. The rapid change in the economic environment made it look as though the antitrust system had become a serious drag on the country's competitiveness when compared to other countries. The area where there was most concern about the negative affect of antitrust on industry was interfirm collaboration. It was widely believed that this was one of the key elements that contributed to Japanese competitiveness, so there was naturally a massive clamour for the US to follow suit. The pressures eventually became so great that legislation, the National Co-operative Research

Act (NCRA), was passed during the first Reagan administration in 1984. Firms that register with the Department of Justice as a consortium for pre-competitive R&D are shielded from treble damages if they are sued for civil antitrust violations.[59] Although this was a major advance, it was felt that the NCRA was inadequate for the task in hand. The protection it offered to US firms from prosecution was nowhere near as comprehensive as that enjoyed by Japanese and European firms, so pressure remained on the government to bring in further legislation to help 'level the playing field' for domestic firms.

To make the US position on interfirm collaboration more equitable with the EU, in particular, Jorde & Teece (1992) proposed a series of amendments to the NCRA.[60] Two of the most important changes they proposed were that protection from antitrust litigation should be extended to manufacturing, and a 'safe harbour' should be defined according to market power for interfirm agreements that involve less than 20–25 per cent of the relevant market.[61] As far as the former point is concerned, this was taken up by the Clinton administration and legislation extending protection to manufacturing was presented to Congress in March 1993. The justification used by Jorde & Teece for these changes was that they expected them to give a vital boost to the country's competitiveness. The theory being that industry would be much freer than it had ever been before to develop collaborative partnerships and therefore would be in a position to increase the overall rate of innovation. As they have stated:

> We have no illusion that our proposed changes, standing alone, would dramatically improve the performance of US industry, although specific industries might be transformed. We recognise that other policies affecting innovation are important such as savings rates, investment in education and technological skills, and appropriate financial and tax incentives. However, the changes we propose in antitrust law are also important and have the attraction that they do not require the expenditure of public funds. In short, we see existing antitrust law as a self imposed impediment to US economic performance.[62]

Jorde & Teece do not claim that their proposed changes 'would dramatically improve the performance of US industry', although they do suggest that 'specific industries might be transformed'. However, whilst they acknowledge that other policies and factors are important, they have completely missed the importance of production. No amount of reform to the antitrust system will rectify the chronic weaknesses which the country suffers from in this area. Antitrust policy was not responsible for the travails of the semiconductor and LCD industries after the late 1970s. It's almost as if the competitive battles, which the US has had with Japan, have only been minor skirmishes. As this is not the case, it is much easier to explain why the country has the trade policy it has.

Trade policy

The US has an equally active trade policy agenda to that of the EU, but where the similarities end is in the style of implementation and their respective agendas. On the whole, the EU has a more subtle approach than the US, which is openly aggressive when it comes to implementing its trade policy. The US has taken an increasingly aggressive stance in recent years as a result of two forces about which the government has been able to do very little: the ballooning trade deficit and the continuing attack on the country's technological leadership. For some inexplicable reason, the government continues to believe that if it has an aggressive trade policy, these problems might somehow sort themselves out. Alas, this has not happened and the consequence has been that the US continues to look abroad for solutions to what are essentially domestic problems.

As far as Tyson (1992) is concerned, the country's trade policy has been designed with two very specific objectives in mind: to improve foreign market access (for US firms) and to counter the effects of targeted programmes of some of its principal trading partners.[63] The way the US has tried to 'open' foreign markets has been to exert considerable pressure, especially on Japan and South Korea which it considers to have 'closed' markets. To make these countries cooperate with it, the US passed some very controversial trade legislation, Section 301 and Super 301 (which has now been scrapped). Failure to comply with the wishes of the US exposes Japan and South Korea to retaliatory action which can consist of restricted access to the US domestic market, or the imposition of extremely high tariffs on their exports.

Tyson openly admits these pieces of legislation 'are forms of aggressive unilateralism'.[64] However, she defends them on the grounds that they are, 'an interim response to foreign trading practices and structural barriers that harm American economic interests and that are not covered by binding multilateral regulations'.[65] While the US obviously has the right to pursue and protect its own national interests, the fundamental flaw with its approach to trade policy is that it is not properly focused. Heavy-handed tactics might work when it comes to prising open what have been deemed to be closed markets, but when the US uses very similar tactics to solve a competitive problem which can only be solved domestically, its trade policy loses credibility.

The anti-dumping petition that was filed against Japanese firms in 1991 is a classic example of trade policy being used to provide relief, in the vain hope that this would help the display industry recover from the pounding it had been getting. When the International Trade Commission (ITC) imposed tariffs on TFT-LCDs and electroluminscent displays, all that it was concerned about was 'punishing' the Japanese firms involved. Apart from the move of notebook production offshore, the other possible long-term effect that the anti-dumping petition is likely to have had, is to make the Japanese firms far less likely to invest in production facilities in the US. Had the petition not been filed in the first place, Sharp might well have built the first production facility in the US rather than an assembly plant.

In light of the petition, and the possibility of more political pressure being exerted on it in the future, it was better for the firm to make some form of investment, albeit of the cheapest kind, than to do nothing at all. At least with some form of operation in the country, Sharp can always say to the government if ever the occasion arises, that the plant can always be upgraded for production as and when the conditions are right. Tyson hits the nail on the head when she says, 'Even with appropriate modifications, trade policy is incapable of solving the competitive difficulties facing this nation's high-technology producers'.[66] The questions that have so far gone unanswered are when will the US propose to modify its trade policy, and when will it appreciate that the action which its display industry has witnessed in the end does the country far more harm than good.

Technology policy

The debate about whether the US needed a technology policy carried on for more than fifteen years before it finally moved to the top of the political agenda and became a major issue in its own right. This was despite the fact that the National Science and Technology, Organisation and Priorities Act, was passed in 1976. It was not until after the election of (former) President Clinton in 1992 that there was any significant movement towards the development of a technology policy aimed at improving the country's competitiveness. Evidence of the

Clinton administration's determination to move ahead on this issue came in February 1993 when it unveiled its ideas for a comprehensive policy.[67] The policy comprised of three main elements, supply-side activities, demand-side activities and the creation of an information and facilities infrastructure. Branscomb (1993) asserts that the major difference between the Clinton administration and the two previous administrations under presidents Reagan and Bush (Sr) was that the latter were only concerned with the supply side, to use technology to achieve federal objectives. Clinton, on the other hand, was keen to embrace all three elements of a technology policy. As Branscomb points out, both Reagan and Bush were loath to adopt a technology policy akin to Clinton's, because they thought it was a smokescreen for an 'industrial policy' which they wanted to have absolutely nothing to do with.

One of the core features of the Clinton administration's bid to develop a successful technology policy was to 'revolutionise' the government's own priorities. The first department to fall victim to the new thinking was the Department of Defense's (DOD) research arm, the Advanced Research Projects Agency (ARPA). In Chapter 9, it is stated that ARPA was previously known as the Defense Advanced Research Projects Agency (DARPA), but it was renamed in 1993, as the administration wanted to emphasise what it intended would be a permanent change in its role. Rather than spending federal resources with only military applications in mind, it was going to focus far more on civilian and 'dual-use' technologies. This was going to be done through the reallocation of existing resources. The distribution of federal R&D resources was to change from 59 per cent military and 41 per cent civilian, to approximately a 50:50 split.[68]

One of the best examples available of Clinton's determination to develop a technology policy is the Flat Panel Display Initiative (FPDI). This was designed to help improve the technological capabilities of US firms to enter the TFT-LCD industry; the emphasis was on a public-private partnership, aimed at helping the private sector to develop its own capabilities to exploit a technology enhanced by public incentives.[69] This was just one of a number of initiatives designed to foster greater collaboration amongst firms, and forge much closer relationships with the government.[70]

With the election of President Bush (Jr) in 2000 the fragile roots of the Clinton's technology policy were abruptly killed off. Following in the tradition of the previous Reagan and Bush administrations, the new administration wanted nothing to do with another form of 'industrial policy'. The Bush administration does have an 'Office of Science and Technology Policy', but its mission is nowhere near the same as the former Clinton administration's was: to reconfigure the institutional structure and policy paradigms developed during the Cold War for the new millennium.

Summary

The preceding analysis has sought to show that the public policy frameworks of Japan, South Korea, the US, and Western Europe are best described as a 'diverse landscape'. For the purposes of stimulating technological innovation, Japan and South Korea have established the most effective frameworks compared to those of the US and Western Europe. The balance which Japan and South Korea have achieved has taken considerable time, and furthermore the frameworks have been constructed with the other elements of their respective NSIs in mind. Compare this approach to the US and Western Europe which have approached the construction of their own public policy frameworks in a more haphazard and piecemeal fashion. The problem for the US and Western Europe is that without an effective public policy framework their own competitiveness will continue to be affected. That aside, Japan and South Korea have their own problems. It must be acknowledged that Japan's own public policy framework needs modification to meet the new challenge it faces from South Korea and a rapidly growing China. South Korea, on the other hand, has to be mindful it does not take its eye off the ball and allow itself to be overtaken by events or it too will find its competitiveness will start to falter under pressure from China, and possibly a resurgent Japan.

Notes

1. Taiwan has been excluded as it is not a member of the OECD and has only been a customs territory of the World Trade Organisation (WTO) since January 2002.
2. Tyson, L (1992) *Who's Bashing Whom: Trade Conflict in High-Technology Industries*, Washington, DC: International Institute for Economics, p. 83.
3. *Ibid.*
4. *Ibid.*
5. Mowery, D & Rosenberg, N (1989) *Technology and the Pursuit of Economic Growth*, Cambridge: Cambridge University Press, p. 274; Jorde, TM & Teece, DJ (1992) 'Innovation, Cooperation and Antitrust' in TM Jorde & DJ Teece (eds) *Antitrust, Innovation and Competitiveness*, New York: Oxford University Press; Holmes, P (1993) 'Competition, Trade and Technology Policy' in M Humbert (ed.) *The Impact of Globalisation on Europe's Industries and Firms*,

London: Pinter; Buigues, P, Jacquemin, A & Sapir, A (eds) (1995) *European Policies on Competition, Trade and Industry: Conflict and Complementarities*, Aldershot: Edward Elgar.

6. *Application of Competition Policy to High-Tech Markets*, (1997) Paris: OECD, GD/(97)44.

7. *Ibid*:7.

8. With the first type of market, current generation products, concern has been voiced particularly by the US, that standard analysis does not take the effect of innovation properly into account. Some analysts have argued that the DOJ/FTC Merger Guidelines (1992) which define markets on the basis of consumers' likely responses to price increases focus on the wrong variable. A critical aspect of competition is product attributes and it is suggested that antitrust markets should be defined with attribute competition in mind. The second type of market, innovation markets, consists of the research and development directed to produce particular new or improved goods or processes and their close substitutes. In defining such markets, an analyst looks to a hypothetical monopolist's ability to profit from reducing efforts in a given line of research and development. The immediate focus of analysis is on research and development activities, but the underlying concern is the ultimate effect in a goods market. The third type of market, future generation products, is essentially about trying to predict the nature of products in the future. This may be difficult because innovations may substantially transform the competitive landscape. Analysts have suggested various techniques for projecting effects in future goods markets based upon observable R&D capabilities and incentives, and the nature of merging firms' existing assets (*Ibid*:7–8).

9. *Ibid*:9.

10. The main source for this section is Matsushita, M (1993) *International Trade and Competition Law in Japan*, Oxford: Oxford University Press.

11. Okimoto, DI (1989) *Between MITI and the Market: Japanese Industrial Policy for High-Technology*, Stanford, California: Stanford University Press, pp. 34–35.

12. *Ibid*:35.

13. Article 2 (6) of the AML defines unreasonable restraint of trade as: 'such business activity, by which enterprises by contract, agreement, or any other concerted activity mutually restrict or conduct their business activities in such a manner as to fix, maintain or increase prices, to limit production, technology, products, facilities, customers, or suppliers, thereby causing, contrary to public interest, a substantial restraint of competition in a particular field of trade'. Matsushita, 1993:136.

14. Kim, L (1997) *Imitation to Innovation: The Dynamics of Korea's Technological Learning*, Boston, Massachusetts: Harvard Business School Press, p. 34.

15. *Ibid*.

16. *Ibid*.

17. *Ibid*.

18. *Competition Policy in OECD Countries 1992–93*, Paris: OECD (1995).

19. Kim, 1997:35. The lifting of the credit controls only applied to the thirty largest chaebols. However, for them to take advantage of this measure they had to reduce internal ownership to less than 20 per cent, raise capital-to-assets ratio above 20 per cent, and offer more than 69 per cent of their shares to the public.

20. *Report by the Government: Summary Extracts for the Republic of Korea,* Press Release 40, Trade Policy Review Board (TPRB), Geneva, Switzerland: World Trade Organisation, October 1996.
21. Press Release 138, Trade Policy Review Board (TPRB), Geneva, Switzerland: World Trade Organisation, September 2000.
22. Kim, 1997:48.
23. *Ibid*:48–49.
24. *Ibid*:50.
25. *Ibid*:65.
26. *Ibid*:49.
27. Holzler, H (1990) 'Merger Control' in P Montagnon (ed.) *European Competition Policy,* London: Royal International Institute for International Affairs, Pinter, pp. 9–10.
28. Gilchrist, J & Deacon, D (1990) 'Curbing Subsidies' in P Montagnon (ed.) *European Competition Policy,* London: Royal Institute of International Affairs, Pinter, pp. 31–51.
29. *Ibid*:32.
30. *Ibid.*
31. *Ibid*:39.
32. Quoted in Buigues *et al.,* 1995:93.
33. Sharp, M & Pavitt, K (1993) 'Technology Policy in the 1990s: Old Trends and New Realities', *Journal of Common Market Studies,* June, Vol. 31, No. 2, p. 143.
34. Private communication.
35. Sharp, M (1991) 'The Single Market and European Technology Policies', in C Freeman, M Sharp & W Walker (eds) *Technology and the Future of Europe: Global Competition and the Environment,* London: Pinter, p. 64.
36. Buiges *et al.,* 1995:93.
37. Quoted in *Ibid*:97.
38. *Ibid*:125–159.
39. Sharp & Pavitt, 1993:135.
40. Sharp, 1991:64.
41. *Ibid.*
42. *Ibid.*
43. *Ibid.*
44. *Ibid.*
45. *Ibid.*
46. *Ibid*:65.
47. Quoted in *Ibid*:59.
48. *Ibid*:61.
49. Neale, AD (1970) *Antitrust Laws of the United States of America: A Study of Competition Enforced by Law,* Cambridge: The National Institute of Economic and Social Research, Cambridge University Press.
50. Bork, RH (1978) *The Antitrust Paradox: A Policy at War With Itself,* New York: Basic Books.
51. *Ibid*:5.
52. *Ibid*:7.
53. *Ibid*:166.
54. *Ibid.*

55. *Ibid*:164.
56. *Ibid*:211.
57. *Ibid*.
58. *Ibid*:4.
59. Branscomb, LM (ed.) (1993) *Empowering Technology: Implementing a US Strategy*, Massachusetts: MIT Press, p. 24.
60. Jorde, TM & Teece, DJ 'Innovation, Co-operation, and Antitrust' in TM Jorde & DJ Teece (eds) (1992) *Antitrust, Innovation and Competitiveness*, New York: Oxford University Press, p. 59.
61. *Ibid*.
62. *Ibid*:48.
63. Tyson, 1992:295.
64. *Ibid*:258.
65. *Ibid*.
66. *Ibid*:286.
67. Branscomb, 1993:1,9.
68. *Ibid*:16.
69. *Ibid*:269.
70. For a more detailed account of these initiatives see *Ibid*:270.

4
Semiconductors: A Truly Global Industry

This chapter analyses the development and progress of the semiconductor industries of the US, Japan, and Western Europe, up to the early 2000s. The role of public policy has been widely credited with the successful development of the US and Japanese industries, and the failure to revitalise the European one. However, the long-term survival of the respective industries is ultimately decided by national firms' ability to keep up with changes in the direction and rate of technological change, how they manage to exploit present and emerging opportunities, and their ability to withstand the frequent and wild fluctuations in supply and demand.

As stated earlier in Chapter 1, there are close parallels between the LCD and semiconductor industries that cannot be ignored. The most immediate one is that the US was the first country to establish an early lead in the two industries. Unlike in LCDs, where it 'gave up' its lead at a very early stage, in semiconductors the US took full advantage of its early lead. As we shall see, it was a mixture of US firms' ability to keep up with changes in the direction and rate of technological change, exploit emerging opportunities, and withstand the frequent and wild fluctuations in supply and demand, which enabled the US to maintain its lead for so long.

It was in the mid to late 1970s when signs emerged that the US was losing its momentum in the industry and it could no longer take its dominance for granted. It was the very same factors that had enabled the US to establish its lead that caused it to experience major problems, and permitted Japan to catch up after approximately two decades. Especially during the 1980s, it was Japan that proved to the world it was much better at managing the factors listed above than the US, and subsequently forged ahead in the industry. Unlike the US and Japan, which have become global players in the industry, Western Europe has never kept pace with changes in the direction and rate of technological change, nor for that matter, coped very well with the frequent and wild fluctuations in supply and demand. Its failure to manage these factors has meant that Western Europe has developed a reputation for being the 'laggard' of the semiconductor industry. It is a trend that can be traced back over more than three decades.

These, and other changes which have hit the semiconductor industry at various points in time over recent decades, provide a number of useful insights that aid our understanding of the development of the LCD industry. These include, amongst others, the role and 'effectiveness' of public policy, the importance of keeping up with the direction and rate of technological change, and the long-term implications which severe swings in supply and demand can have. Throughout the semiconductor industry's history a number of different factors have come into play. They are still very much at work and continue to influence the ever changing character of the semiconductor industry. Therefore, it makes sense to improve our understanding of the interrelationships between them, and how they have helped shaped the semiconductor industry into what it is today.

United States

The foundations of the US semiconductor industry were laid with the invention of the transistor by the Bell Laboratories in 1947. However, it was not until the early 1950s that the number of US firms producing transistors began rapidly to increase. In 1952 there were eight firms producing transistors, a year later, there were fifteen firms, and by 1956, the figure had risen to twenty-six. The industry included firms like General Electric, Westinghouse, Sylvania, Raytheon, CBS, Tung-Sol, and RCA.[1] As late as 1959, this group of eight firms and Bell were contributing 57 per cent of the total R&D expenditure on semiconductors and 63 per cent of major innovations, most of the latter coming from Bell.[2]

Like the vast majority of inventions, Braun & MacDonald (1980) explain that before the transistor could be transformed from a scientific curiosity into a commercially viable product it had to undergo extensive development.[3] Sufficient development work had been carried out on the transistor by October 1951, that Western Electric, the production division of AT&T, was able to commence production of transistors on a commercial basis. Two of the earliest applications for transistors were hearing aids and telecommunications.

By the mid-1950s, transistors were being used in the production of household radios, car radios and computers. The use of the transistor in the production of computers had enormous attractions to firms that wanted to develop and advance in the industry. Until the advent of the transistor, valves had been used to produce computers, but compared to the transistor, the valve was immensely cumbersome and inefficient. A computer made of valves was extremely large, it consumed an awful lot of power and generated a great deal of heat. The advantage of the transistor over the valve is immediately apparent. In 1955 IBM marketed an early version of a commercial computer which replaced 1,250 hot valves with 2,200 transistors.[4] Size was reduced, the need for cooling removed, and power consumption reduced by 95 per cent.[5]

One of the most interesting and valuable aspects of Braun & MacDonald's analysis is how many of the large established electronics firms and their top management were totally unaware of the revolutionary nature of the transistor. The semiconductor department was more often than not little more than an offshoot of the valve division. The lack of attention to semiconductor technology generally, meant that those individuals who were deeply immersed in its development and appreciated its enormous value, were prone to switch from one firm to another in search of ever greener pastures, and of course, the lure of ever higher salaries. This high rate of labour mobility has remained a distinctive characteristic of the US industry ever since, and has been a major source of entrepreneurial individuals who have been eager to set up their own firms to exploit opportunities they have seen emerging over the horizon.

At the same time as the semiconductor industry was expanding, so too was the electronics industry. The growth of the electronics industry between 1950–56 can be seen in Figure 4.1.

In 1950–51 the main driving force behind the growth of the electronics industry was the consumer, by a considerable margin. By 1953, however, the pendulum had swung in the completely opposite direction. The consumer had been replaced as the major source of demand by the military. In the accounts of the US semiconductor industry by Braun & MacDonald (1980) and Tyson (1992),[6] a major discrepancy exists over whether or not it was civilian or military considerations which were the source of the industry's rapid growth from the beginning of the 1950s. Both remark that the transistor was invented by the Bell Laboratories. However, as noted above, Braun & MacDonald convincingly argue the transistor was a civilian invention and could never

Year	Total sales ($m)	Government (Per cent)	Industrial (Per cent)	Consumer (Per cent)	Replacement Components (Per cent)
1950	2705	24	13	55.5	7.5
1951	3313	36	13.6	42.3	8.1
1952	5210	59.5	9.5	25	6
1953	5600	57.7	10.7	25	6.6
1954	5620	55.2	11.5	25	8.3
1955	6107	54.6	12.2	24.6	8.6
1956	6715	53.5	14.2	23.8	8.5

Figure 4.1 Uncorrected value of US electronics sales by end-use category 1950–56
Source: Braun & MacDonald (1980), p. 79

be conceived in any way as a military achievement.[7] Tyson, on the other hand, has claimed the industry 'was born out of the US effort to develop ever more reliable and sophisticated military equipment, and indeed its military significance has never been doubted'.[8] Where the two accounts do converge is over the role of the military as a major source of demand and as a critical source of R&D funding during some of the crucial stages of the industry's development.

Figure 4.1 shows that in 1952 the military was the biggest end-user of the electronics industry by a considerable margin. It accounted for 59.5 per cent of the electronics industry's sales, whilst the consumer accounted for only 25 per cent. In the same year, out of 90,000 transistors produced, the military bought nearly all of them.[9]

Military funding for the development of the transistor commenced in 1951 when the army, navy and air force, gave the responsibility for overseeing its development to the Signal Corps. The Signal Corps used what were known as production-engineering measures (PEMs) as a means of defining and achieving its key objectives. These were to reduce the cost of the transistor, improve its reliability and performance, and to insure the semiconductor industry built and maintained enough production capacity.[10] When it came to actual R&D funding for semiconductors, the military spent comparatively little, especially during the 1950s. Prior to 1956, the Signal Corps spent less than $500,000 in this area, and about $1 million after this date.[11] The

Year	Diodes & rectifiers (M)	Transistors (M)	Integrated circuits (M)	Discrete equivalents of integrated circuits (M)	Total discrete components or equivalents (M)	Percentage of components in integrated circuits
1963	627.1	302.9	4.5	108	1038	10.4
1964	762.1	398	13.8	331.2	1491.3	22.2
1965	1072.6	631.7	95.4	3052.8	4757.1	64.2
1966	1520.8	877.3	165	5280	7678.1	68.8
1967	1461.1	792.1	1967	8582.4	10835.6	79.2
1968	1618.8	951.7	247.3	11840.4	14410.9	82.2
1969	2099.7	1249.1	423.6	23721.6	27074.4	87.6
1970	1866	976.9	490.2	27451.2	30294.1	90.6
1971	1473.1	880.7	635.2	40652.8	43006.6	94.5

Figure 4.2   Shipments of discrete components and integrated circuits 1963–71
Source: Braun & MacDonald (1980), p. 112

Signal Corps preferred to channel funds into the areas mentioned above and the pattern of funding duly reflected this. PEM support is estimated to have totalled around $50 million between 1952 and 1964.[12] Initially, it was the large established electronics firms which benefited disproportionately from government funds. In 1959 Western Electric and the eight valve firms received 78 per cent of R&D funds from the government, even though they controlled only 37 per cent of the market.[13]

By the end of the 1950s, the US semiconductor industry had come a very long way from its humble beginnings a decade earlier. Throughout the period, the industry enjoyed enormous growth, stimulated by demand and financial support from the military. The downside to this era of growth, which no firm liked, but nevertheless had to accept, was that as production rose, the price of devices plummeted. One area that experienced a dramatic fall in prices were silicon transistors. In 1957 one million silicon transistors were sold by US industry at an average price of $17.81 each. By 1961 annual sales had risen to 13 million, but the average price of each silicon transistor had declined to $7.48. This trend continued until the middle of the decade. By 1965 sales of silicon transistors had reached 274.5 million, but the average price had fallen to $0.86 each. The relationship between rises in the level of production and dramatic reductions in price, which first became apparent at the end of the 1950s and early 1960s, has since become the bane of the semiconductor industry.[14]

At the beginning of the 1960s the industry experienced its first major technological 'revolution' since the invention of the transistor, with the appearance of the integrated circuit (IC). And with it came the predictable shakeout amongst the firms as they struggled to get to grips with this new technology. The changes this new technology brought about helped in turn to change the way the government allocated the financial support it gave to firms.

Texas Instruments was the first firm to file a patent for an IC in February 1959,[15] but the first recorded commercial use of an IC, in a hearing aid, was not until December 1963.[16] The appearance of the IC on the market did not lead to a rapid rise in production levels. As shown in Figure 4.2, the production of transistors far outstripped the production of ICs throughout the 1960s.

The biggest hurdle faced by firms in the first half of the decade in the production of ICs was yield, that is, the level of output totally free of defects. The trouble was that the technology was moving rapidly towards miniaturisation and greater integration, which took firms time to adjust to. When firms decided to start to manufacture ICs, one aspect of their production, which they did not have to worry about too much, was the production process. The same process used to produce transistors could also be used to produce ICs. What caught many firms out, however, was the increase in the number of manufacturing steps it took to make an IC. Whereas it took approximately fourteen steps to make a transistor, it required approximately twenty-two to make an IC, an increase of about a third in the number of steps required.[17] The effect on yield was dramatic. A 95 per cent yield for each step would produce transistors at 50 per cent yield, but ICs at only 25 per cent yield.[18]

High prices for the early ICs was perhaps the most immediate knock-on effect of low yields. As shown in Figure 4.3, in 1962 the price for an IC was $50, but a year later the price had declined to $31.60, a decline of 38 per cent as yields improved. At these levels, the IC was still an expensive commodity, but firms had little trouble finding customers for them. Because of their size and reliability, once again the military proved a captive market. Between 1962–65 the military was the semiconductor industry's biggest customer. In 1962 it absorbed the industry's entire output of ICs. Such was the demand from the military that by 1965, it still absorbed just under three-quarters of the industry's output. After that date the importance of the military as a customer gradually diminished. In view of the scale of the military's purchases of ICs, it is easy to get the misleading impression from Figure 4.3 that the military was responsible for purchasing a high percentage of the electronics industry's output, especially from the early to mid-1960s. This was not the case as Figure 4.3 clearly shows. The electronics industry produced a number of items other than just ICs, in vast quantities, for which the military had little or no requirement.

The abundance of opportunities in the semiconductor industry encouraged a rash of new firms to enter the industry helped by the switch in funding away from established firms to highly innovative ones such as Motorola and Texas Instruments. According to Borrus *et al.* (1983) 30 new merchant firms were formed by individual entrepreneurs between 1966 and 1972.[19] The formation of new firms was greatly facilitated by the ease of capital available from the venture capital market eager to profit from the growth of the industry. Competition between firms sustained the momentum of innovation at a high level and the industry at the forefront of technology. The pace of innovation, amongst other things, reduced the lead-time and improved the reliability of new devices and generated spillovers with civilian applications.[20] The other important source of demand for semiconductors, of course, was the growing computer industry. It was

Year	Total production ($ million)	Defence procurement as a percentage of total	*Price ($)
1962	4	100	50
1963	16	94	31.60
1964	41	85	18.50
1965	79	72	8.33
1966	148	53	5.05
1968	228	43	3.32
1967	312	37	2.33

Figure 4.3   Government procurement of integrated circuit production 1962–68
Source: Okimoto *et al.* (1984), p. 85 & * Braun & MacDonald (1980), p. 113. Prices derived from Table 8.2

an additional stimulus to the semiconductor industry, which effectively put it under continuous pressure to improve the quality and performance of its products by the setting of high standards.[21]

The vibrancy of the industry and lack of any serious competition presented the industry with a window of opportunity to expand overseas. In the decade between 1964–74, US firms were able to penetrate national markets in Europe, Latin America and Asia without encountering any serious problems. The only country which made a concerted effort to block US firms' access to its national market was Japan. The main incentive for US firms to shift production offshore was the lure of cheap labour as semiconductor assembly was then a labour intensive process. Part of the reason why the number of US offshore assembly plants had reached 69 by 1974[22] was because the US industry assumed the extensive use of cheap labour was one of the best ways of maintaining its competitive advantage.

The mid-1970s was the high point for the US industry. It could not only look back over the preceding years with considerable satisfaction at its innovative performance, it had expanded aggressively overseas and in the modern sense of the term, had become a truly global industry. Superficially at least, the US industry appeared to have little to worry about. Lurking under the surface, however, were structural deficiencies, which until then had not manifested themselves. There was a glimpse of at least one of the most serious deficiencies in the aftermath of the recession that struck the semiconductor industry during 1974–75. Due to the lack of sufficient capital, US firms cut back on investment and production capacity in reaction to the poor business conditions which were prevalent in the industry at the time. When demand recovered in 1976–77, the US industry found itself short of capacity and some of its former customers had to look elsewhere for their supplies. While the recession of the mid-1970s certainly gave the US industry a jolt, it was the recession which struck the industry in 1985–86, which brought out into the open the striking differences between the US and Japanese industries, and led to a permanent change in the position of the two countries in the industry.

When the US industry 'stumbled' in the mid-1970s it created a huge vacuum. That, of course, was promptly filled by the US's one and only major competitor, Japan. Having been given this opportunity, Japanese firms duly went about exploiting it for all it was worth. Within three years, 1976–79, Japan's exports of ICs had increased from $76.7 million to $494.5 million, an increase of 644.8 per cent![23]

To make matters worse for the US industry, evidence emerged that Japanese semiconductors were superior in quality to American ones.[24] Both Okimoto *et al.* (1984) and Forester (1993) give the same example, which received widespread publicity at the time, and completely stunned the US industry. In a presentation at a conference in March 1980, a senior executive of Hewlett Packard revealed evidence that showed Japanese 16K DRAMs were of a consistently higher quality and more reliable than US 16K DRAMs. The firm had been importing Japanese DRAMs since 1977 as a result of the 'damage' the US industry had inflicted upon itself during the recession of 1974–75.

The cutbacks made by the US industry meant that when demand recovered in 1976–77, it was unable to meet Hewlett-Packard's demands. In 1979 the firm carried out tests on 300,000 DRAMs,[25] half from three US firms, and the other half from three Japanese firms.[26] The statistic that everyone found so shocking was that Hewlett-Packard had found that the best US firm had a failure rate six times that of the best quality Japanese firm.[27] Faced with the prospect that unless it acted quickly, the 'quality gap' could get even worse, the US industry took immediate action to deal with the problem. Its quick response meant that by the mid-1980s, the 'quality gap' was no longer the concern it had been.[28] In this area at least, the US industry showed its Japanese counterpart it too had the capacity to respond quickly and effectively when presented with a major challenge.

Japan

The emergence of the Japanese industry as a major competitive force has proved to be no aberration and has continued to be a source of friction between Japan and the US. As mentioned earlier, public policy has played a crucial role in creating the foundations from which the industry grew, but the government did not have to start from scratch to develop the industry as the seeds of one already existed. One of the first steps the government took to get the industry off the ground was in 1957, when it passed the 'Electronics Industry Special Measures Promotion Law' which gave it all the necessary powers to develop the semiconductor industry. Amongst other things, these included the power to direct electronics R&D, to provide subsidies and ensure bank loans to producers, and to set production quantity and cost targets.[29] The industry was singled out for special support on the grounds that it fulfilled the following MITI criteria: income elasticity, that is, the demand for the output of the industry must increase more than the average for the economy as a whole as income rises; and productivity must increase more rapidly than the average.[30]

The roots of the industry were, so to speak, established very early on as the divisions of the large vertically integrated electronics firms started making semiconductors soon after the discovery of the transistor. Unlike in the US where defence procurement and the computer industry were the key sources of demand, initially it was the consumer electronics industry that absorbed the bulk of domestic production (60 per cent of production in 1968),[31] the national telephone carrier, NTT, and then later on the computer industry. While the consumer electronics industry was a stable source of demand on which the semiconductor industry could depend, it was NTT that put the issue of quality at the top of the agenda. In the early 1970s, NTT required large amounts of 16K DRAMs, and it established quality standards that were very high, even for Japan, and these quickly became the industry standard. NTT published quarterly reports for its suppliers to keep them informed of the performance of their product, and consequently firms took every precaution to adhere to the very rigid specifications just to maintain a small market share. If firms failed to adhere to the standards demanded by NTT and at the same time provide a low cost product, they would simply be forced out of the fiercely competitive domestic market.[32]

In addition, the Japanese government strictly monitored foreign firms' access to the Japanese market. As US firms found to their cost, they were prevented from establishing any real presence in Japan by a range of formal and informal barriers. The only way they managed to establish any presence at all was by succumbing to pressure from the Ministry of International Trade and Industry (MITI) to sell their patents and know-how. Together, these measures meant that US firms had only managed to capture a 20 per cent share of the Japanese IC market by 1975.[33] Apart from the highly protectionist measures that the Japanese firms obviously benefited from, the government also successfully managed to vastly improve the technological capabilities of firms. These were raised through various government-sponsored research programmes. After 1970 a raft of programmes were launched in areas such as electron-beam exposure, basic materials research, large-scale integration (LSI) production and discrete devices. The most successful of all was the very large scale integration (VLSI) programme.[34]

The improved technological capabilities of the firms enabled the industry generally to move a lot faster along the learning curve, and the progression up the technological ladder was consequently less arduous. However, there was only a very short interlude before clear signals emerged of the progress made. The success the firms had enjoyed in 16K DRAMs was replicated in the next generation, the 64K DRAM. The major difference between the two generations was that the latter was the more sophisticated and powerful DRAM.[35] Forester argues it was analogous to a blackboard. Information stored on it could be erased from it, making it ready for reuse. In the early 1980s, the 64K DRAM became the standard semiconductor used in a number of important applications such as electronics products, telecommunication systems and computers.[36] For the respective countries competing in the market there was much at stake. The country that came out on top stood to make significant profits and generate huge exports. An indication of how much the market was worth can be seen from the value of Japan's IC exports between 1981–83 – $3,833 billion alone.[37] This rapid rise in exports, which incidentally had started between 1978–79, meant that by the end of 1981 Japan had equalled the US in terms of market share.[38]

The competitive position of the Japanese semiconductor industry received an additional stimulus by its choice to opt for CMOS (complementary metal-oxide semiconductor) technology rather than NMOS technology. Again, the accounts of how and why CMOS triumphed over NMOS technology differ enormously. Either way, the conclusions are the same, the US was put at a major disadvantage due to its long 'association' with NMOS technology. Tyson (1992) maintains it was the result of an innovation in lithography equipment that promptly led CMOS technology to become the low cost technology by 1983–84. Prior to this innovation NMOS technology had been the low cost technology. She goes on to say that most US firms had used NMOS technology since the 1960s in the production of DRAMs, especially those DRAMs pioneered by Intel.[39] The Japanese semiconductor industry, on the other hand, had decided in the early 1970s not to use the same technology as the US semiconductor industry. The most probable explanation for this is that NMOS technology was not suitable for widespread use in the consumer electronics industry. There was no point in the Japanese industry using a technology that was used predominantly in the manufacture of computers and other military and industrial applications. The Japanese semiconductor industry therefore had to develop a technology that was suitable for the industries it served.

Forester infers that the US semiconductor industry was 'content' to stay with NMOS technology because of its long familiarity with it. He claims CMOS technology was simply a much more energy efficient technology than NMOS technology was. As CMOS semiconductors only used 10 per cent of the power of NMOS semiconductors, Japanese firms were able rapidly to increase exports of battery-operated digital watches, calculators and other consumer products that contained CMOS semiconductors.[40] There is more than a grain of truth to this claim. In Chapter 7, it is pointed out that the Japanese government came under increasing pressure from the US government to limit its exports of such items during the early and mid-1970s.

The move from the 16K DRAM to 64K DRAM turned out to be an important watershed for the semiconductor industry as a whole. Firstly, during the transition the number of US firms active in the sector fell precipitously. Tyson (1992) states that seven out of nine US firms were forced to make an exit from the industry.[41] The only remaining firms were Texas Instruments and Micron Technology. As with every other recession, the 1985–86 recession saw prices take a plunge. In April 1984, a standard 64K DRAM sold at $3.50, but a year later the price had dropped to $0.75. Prices reached a low point of just $0.30.[42] This is a startling drop considering that back in 1980 the price of a 64K DRAM was $100.[43] The financial losses incurred by US firms were so bad that many had no option other than to vacate the industry.

Secondly, the contraction of the US industry allowed Japanese firms to capture an increasing share of global semiconductor revenues. By 1986 Japanese firms had captured 46 per cent of global revenues compared to the US firms' share of 40 per cent.[44] This rise in Japan's share of global revenues was a reflection of the country's increasing stranglehold on the DRAM industry that had gradually been getting stronger since the late 1970s. By 1981, for example, Japanese firms had a 70 per cent share of the global market in 64K DRAMs, before seeing it fall back to 54 per cent in 1984.[45] In the next generation of DRAMs, the 256K DRAM, Japanese firms did even better than they had with the 64K DRAM. In 1982 Japanese firms commanded an astonishing 92 per cent of global market in 256K DRAMS. Within three years, even that remarkable performance had been eclipsed when in 1985 they captured 96 per cent of the global market in 1M DRAMs.[46]

Thirdly, the structural deficiencies in the US industry were exposed. In contrast to the Japanese industry, which is relatively homogeneous, made up of large vertically integrated firms, the US industry is diverse, consisting of vertically integrated firms, captive producers, and independent merchant producers.[47] The exit of so many US firms in one go showed that the structure of US industry itself was under intense strain from the financial pressures. It was not just the constant downward pressure on prices that firms had to cope with. They had to contend with the additional burden of the escalating costs of building fabrication plants. In the past twenty-five years or so, costs have risen remorselessly, to the point where even the largest of firms are having problems raising the capital required to build the latest generation of fabrication plants (or 'fabs'). In 1970 the cost of building a fabrication plant was $3 million, but by 1980 the cost had reached approximately $75 million.[48] Throughout the 1980s, costs carried on rising. By 1985 the cost of a fabrication plant had risen to around $150 million, and by 1991, costs had shot up to approximately $500 million.[49] And by the late 1990s, the cost of a new fabrication plant had more than doubled to well in excess of $1 billion.[50] While the US industry was under intense financial pressures, the Japanese industry, on the other hand, was in a much better position to handle the situation. Because of its structure and ties to the *keiretsu*, Japanese firms were better able to shoulder sustained losses over a lengthy period of time, and at the same time continue investing in R&D and production capacity.

By the mid-1980s, there was huge political pressure in the US to reach some agreement with Japan to prevent any further decline of the US industry. The environment in which an agreement was eventually reached was far from ideal. The main bone of contention between the US and Japanese governments was that no reciprocal relationship existed between the respective industries. Within the US semiconductor industry there was a widespread belief that its troubles were being compounded by its 'lack of access' to the Japanese market. US firms had very strong suspicions that their access to the Japanese market was being limited by 'unfair trading practices', which was detrimental to the health of the US industry. To US firms, this was grossly unfair as Japanese semiconductor firms not only had 'easy access' to the US market, they were also guilty of 'dumping' semiconductors on the US market.

In response to the problems facing the US firms, the first 5-year Semiconductor Trade Agreement (SCTA) was concluded between the Japanese and US governments to 'manage' the industry in August 1986. The SCTA stipulated that US firms' access to the Japanese market would be improved, which included a declared intention to boost the foreign share of the Japanese market to 20 per cent within five years, and that Japanese firms would stop dumping semiconductors on the

US market.[51] To prevent future dumping the US government established 'foreign market values' for each Japanese firm.[52] On the basis of the cost data supplied by each individual firm, the US government could determine what the average production costs were of a particular firm, and see for itself how costs varied within the Japanese industry. Armed with this information the US government effectively set a 'minimum price', that is, a foreign market value for the sale of DRAMs on the US market for each firm concerned. Under this criterion, a Japanese firm would be guilty of dumping whenever the prices of its DRAMs fell below its average cost price, plus an 8 per cent profit margin which is allowable under US anti-dumping laws.[53] This framework severely restricted price competition amongst firms. Tyson notes that a Japanese firm was 'free' to cut its prices below those of any other Japanese or US firm, but only to the level of its 'foreign market value'.[54] Under the SCTA, price levels were jointly monitored by the US Department of Commerce and MITI, which maintained a very close watch on Japanese firms' export prices.[55]

Tyson highlights three distinctive features of the SCTA which were going to act as the linchpins for managing the industry. Firstly, costs and prices for other semiconductors, not just DRAMs, were to be monitored in anticipation of 'potential' anti-dumping investigations in the future. Although dumping of the products concerned had not actually occurred, their continual monitoring was meant to pre-empt any attempts to do so. The 'logic' behind this was to stop US firms from being 'irreparably' damaged. US firms had complained that because an anti-dumping investigation is a long winded process, by the time it has been completed and duties are imposed (where applicable), they have already by then suffered extensive damage.[56]

Secondly, as it concerned the two main competitors in the semiconductor industry, the SCTA had to be structured to take account of the global nature of competition. As mentioned above, one of the two prime objectives of the SCTA was to bring to a halt the dumping of semiconductors on the US market by Japanese firms.[57] The situation which the US government, in particular, wanted to avoid at all costs were prices in its own domestic market becoming artificially inflated at the expense of consumers. As US firms faced very stiff competition from Japanese firms in world markets, there was a very strong possibility that prices could tumble elsewhere, but not in the US. US consumers would then have the opportunity of buying their supplies of semiconductors where they could obtain them at the lowest possible price.

Thirdly, the anti-dumping measures contained in the SCTA focused on costs and pricing behaviour of individual firms. This was an explicit admission that Japanese firms were not all the same. There were high-cost producers as well as low-cost ones. As the two types competed with each other, costs would fall, therefore giving low-cost US firms an opportunity to increase their market share. The theory was that combined with the 20 per cent 'target' to increase the foreign share of the Japanese market, high-cost Japanese firms would be much less inclined to embark on further capacity expansion.[58]

Shortly after the ink had had time to dry on the SCTA, US firms adopted a number of measures to increase their market share in Japan between 1986–90.[59] Amongst other things they increased the number of sales offices by 30, in addition to the 42 offices that they already had; they opened another 16 new design centres along with 6 new testing and quality centres (which brought the total to 18); and they invested in 4 new failure analysis centres.[60] As a means of boosting their market share in Japan, increasing the number of sales offices and opening various types of centres was one of the cheapest available options open to US firms. The much more expensive option of investing in fabrication plants in Japan, the vast majority of US firms decided not to opt for. The justification for this seems to have been that the initial up-front investment did not justify the short-term sales projections.[61] Tyson notes that only four firms had some form of fabrication plants in Japan. They were LSI Logic, which ran a front end plant with Kawasaki Steel; Analog Devices, which operated an assembly and test plant; Texas Instruments which operated three wafer fabrication, assembly, and test complexes; and Motorola, which had a wholly owned integrated facility and also a joint venture with Toshiba.[62]

There is certainly some merit in the proposition that the main reason why the great majority of US firms did not decide to invest in fabrication plants was due to financial considerations. However, there are grounds for suspecting that there was more to this than meets the eye. The telltale sign that something more fundamental was wrong is the joint venture between Motorola and Toshiba. The agreement, like many others before it, involved a swapping of technologies between the respective parties. Motorola agreed with Toshiba it would swap its technology in 8, 16, and 32-bit microprocessor technology in return for Toshiba's DRAM technology.[63] Once in possession of this DRAM technology, Motorola was then able to proceed with the production of 1M DRAMs in Japan and Scotland.[64] Revealingly, for Motorola this was one of the few ways it could re-enter the DRAM market, which it had vacated in 1985.[65]

The most worrying aspect for US firms about this particular joint venture is that if a firm like Motorola, which has a highly innovative record, can 'fall behind' in DRAM technology, it is perfectly possible for lesser firms to do so as well. As the example of Motorola has illustrated, a firm can only make a successful re-entry into a fast changing industry when it is in the possession of the appropriate technology and can produce it on a commercially viable basis. Even in cases where firms were still very active in the industry, their apparent reluctance to compete with Japanese firms in *producing* DRAMs signals that there were more than financial considerations at work. Surely, the most appropriate way for US firms to increase their market share over the long term was to strengthen both their R&D and production activities together, just like Japanese firms have been doing for years. It does not seem to make much sense to focus on one end of the spectrum when your major competitor is focusing on both. Arguably, investing resources in one area and not the other could be construed as being an implicit admission that the area that is being denied resources has become a lost cause.

In many ways the action taken by US firms to improve their share of the Japanese market was only going to be effective up to a certain point. Under the provisions of the SCTA, it would appear, it was envisioned that US firms would increase their market share not only by ensuring that the quality of DRAMs they supplied were of roughly the same standard as Japanese ones, but they would readily invest in additional production capacity in Japan to achieve greater economies of scale. By offering DRAMs at a similar price and quality to Japanese ones, US firms should then have had little trouble in persuading Japanese firms to increase their purchases of American DRAMs. At least in this way, US firms could say that they had increased their market share in a 'bona fide' manner. Any increase in market share would be down to their enhanced competitiveness rather than additional 'assistance' from the US government.

By not investing in additional production capacity in Japan, the majority of US firms look as though they were pinning their hopes on increasing their market share in one of two ways. Either they expected to meet the increase in demand for DRAMs from their facilities located outside of Japan, or they, expected the US government to exert pressure on the Japanese government to open up its market sooner rather than later. The overwhelming evidence suggests that US firms were hedging their bets on the latter. How could the US industry hope to compete successfully against the Japanese industry in DRAMs after so many US firms had exited the industry in 1985? Behind the 'facade' of the SCTA being a 'temporary' measure for US firms to re-establish a presence in DRAMs, the objective of US firms from the start has been to use it as a mechanism to increase their sales in Japan of other types of semiconductors where they are much more competitive.

As might be expected, without a healthy DRAM industry, it was unrealistic for the US government to expect the country's share of the Japanese market to rise automatically soon after the SCTA had been concluded. However, the US government, perhaps not fully aware of how far the domestic DRAM industry had actually fallen, imposed 100 per cent tariffs on carefully targeted, Japanese-made computers and electropneumatic hammers in April 1987, in response to Japan not adhering to the terms of the SCTA.[66] The US government's swift action was precipitated by the fact that seven months after the SCTA had been signed, US firms' share of the Japanese market had not increased. Moreover, the Semiconductor Industry Association (SIA), the trade body of the US semiconductor industry, obtained documentary evidence that showed Japanese firms were selling DRAMs in markets in the Far East at below their 'foreign market values'.[67] The US trade sanctions had the predictable effect on the Japanese electronics firms. Rather than run the risk of more punitive sanctions being imposed, they made a concerted effort to increase their purchases of US semiconductors. This helped drive the foreign market share to approximately 14.3 per cent by mid-1991, its highest level ever.[68]

How far the SCTA actually affected prices is a matter of some debate. An element of caution is needed here as Tyson provides contradictory evidence that, on the one hand, suggests the SCTA had a 'positive' influence, and on the other, that it did not. From the time the SCTA came into force until 1989, DRAM prices were at times 'allegedly' very erratic. However, these fluctuations do not show up in Tyson's own data for the 256K DRAM and 1M DRAM between 1986–89.[69] In the second quarter of 1986, the average price for a 256K DRAM was $2.50, compared with $3.50 in the first quarter of 1989. There was only one quarter during this whole period when prices rose above this level and that was in the last quarter of 1988. When they did rise, they only did so very slightly. At their 'high' point the price of a 256K DRAM was only $3.50.[70] From the third quarter of 1989, prices for the 256K DRAM were consistently below the $3 level.[71] As with the 256K DRAM, prices for the 1M DRAM exhibited no major fluctuations during the period in question. After falling rapidly from an average price of $45 in the third quarter of 1986 to $20 in the first quarter of 1987, prices remained within the $16–18 range for the next two years.[72]

This apparent price stability masked fluctuations experienced in specific markets. Tyson claims the US experienced at least two periods when prices fluctuated wildly. The first was in the first few months after the conclusion of the SCTA when US customers reported the prices of 256K DRAMs (imported from Japan) increased between two and eight times their pre-agreement price,[73] and the second was in 1988 when spot prices for 256K DRAMs tripled over a four-month period.[74] In Japan, the prices of the 256K DRAM and 1M DRAM were noticeably lower than in the US, Europe, and East Asia from the time of the SCTA through to the middle of 1989 when these differentials evaporated due to the introduction of the 4M DRAM.[75]

Looking at these price trends, there is a strong inclination to reach the conclusion that these erratic fluctuations were directly attributable to the SCTA. However, this would be a bad mistake to make. Both before and after August 1986, there were a number of developments which by themselves would have had a far greater impact on the behaviour of prices than the SCTA could ever have done. Prior to August 1986, the biggest influence on prices by a long way was the direct action taken by Japanese firms to bolster DRAM prices, after the exit of a number of US firms. In an article published in the highly respected, *Nihon Keizai Shimbun*, in January 1986, it was reported that in the previous month most Japanese firms had announced a 20 per cent across-the-board price increase for 64K DRAMs and 256K DRAMs.[76] In another article, in the same month, Japan's two largest producers of 256K DRAMs, NEC and Hitachi were reported to have been in the process of cutting production because of massive inventories.[77]

After August 1986, the main developments to influence DRAM prices were the recovery in demand in the computer industry in 1987 and problems associated with bringing the 1M DRAM into production.[78] Finally, if there really was a direct correlation between DRAM prices and the SCTA, then the fluctuations seen in DRAM prices should theoretically have also have been witnessed in the world prices for EPROMs. The anti-dumping provisions made for DRAMs were not that much different from the ones made for EPROMs, yet the behaviour of DRAM prices differed vastly from EPROM prices.[79]

Since the first SCTA expired, there have been two subsequent ones, but even with these in place the US DRAM industry has continued its steady decline. When the original SCTA expired in July 1991, the Japanese government came under intense pressure from the US government to reach yet another agreement. Although the dumping problem had not gone away, the major stumbling block once again was the thorny issue of foreign market share. In the early part of 1991, the foreign market share of the Japanese market was 'only' 14 per cent.[80] As one of the objectives of the original SCTA had been to increase this figure to 20 per cent, the US government felt justified to argue there was a need for another agreement on the grounds that one of the major conditions of the original agreement had not been met. With little solid evidence of its own which suggested otherwise, the Japanese government was compelled to sign a new five-year agreement with the US government, or face the likely imposition of major trade sanctions.

Because the US government considered the original SCTA a failure, it made sure that in the second agreement the 20 per cent target was deliberately specified. In the course of the negotiations it was agreed by both governments that the target should be realised by the end of 1992.[81] For its part, the Japanese government was very careful to make it crystal clear in the agreement the target was just that. Under no circumstances was it to be interpreted that the Japanese government had agreed to 'hand over' a fixed percentage of its semiconductor market for the benefit of foreign firms, at the behest of the US government. The other major modifications made to the second SCTA were the US government's agreement not to set foreign market values for DRAMs and EPROMs, and its agreement to 'give up' its responsibility for collecting cost and price data from Japanese firms.[82] In future the responsibility for collecting and maintaining all such data would be that of the Japanese firms. In the event of an anti-dumping investigation, Japanese firms would be obliged to provide the US government with all the appropriate information within 14 days of any request.[83]

The willingness of the US government to unilaterally impose sanctions on Japan guaranteed that the Japanese government took the 'necessary steps' so that by the end of 1992, the 20 per cent target had been reached. According to one member of the Japanese semiconductor industry, what this boiled down to in practice was industry officials pleading with firms to buy foreign semiconductors even if it meant that they had to throw them away at a later date.[84] Once the deadline had passed, the sense of urgency which had gripped the Japanese government gradually evaporated, allowing the foreign market share in the first three quarters of 1993 to slip to 18.3 per cent.[85] By allowing the foreign market share to fall below the politically sensitive 20 per cent level, the US government immediately suspected the Japanese government of ignoring its previous commitments and quickly resorted to its age-old tactic of pressurising the Japanese government. This time around the US government urged the Japanese government to adopt an action plan to halt the decline.[86] Precisely what this so-called action plan was to comprise of is unclear, but the speed of the US government response to the fall in market share, and the ever present threat of sanctions gave the Japanese government back its sense of urgency. Knowing full well the US government was in no mood to tolerate any prevarication and meant what it said, Japanese firms rapidly increased their imports of foreign semiconductors so by the fourth quarter of 1993 the foreign market share had increased to 20.7 per cent, its highest level ever.[87]

The effect on the Japanese government of the constant US government pressure became evident when the second SCTA came up for its mid-term review in July 1994. For the first time, the Japanese government started to show its displeasure in public at the US government's attitude over the 'need' for the SCTA, even though the foreign market share had reached an all-time high of 21.9 per cent in the second quarter of 1994.[88] The Japanese government claimed that as the foreign market share was well above the 20 per cent target, the SCTA had served its purpose and should therefore be scrapped. The US government immediately rejected this proposition claiming the SCTA was indispensable, as market access for foreign semiconductors was from its standpoint still not totally satisfactory.[89]

In the course of the following year, the contrasting views of the Japanese and US governments became more pronounced as the expiry date for the second SCTA drew closer. The Japanese government maintained its earlier claim that the SCTA should be scrapped once and for all, and not just because the foreign market share had risen above the 20 per cent level.[90] To the Japanese government, the SCTA had become obsolete in a global industry, characterised by a complex web of alliances and agreements between firms, which had rendered the nationality of semiconductors meaningless. Moreover, the semiconductor industry had changed radically since the mid-1980s. The earlier fears that the US industry would lose its competitive edge against the Japanese industry had dissipated, especially after it was confirmed that Intel had been ranked as the world's number one semiconductor maker.[91]

The response from the US government was predictable as ever. Rather than making any attempt to counter these 'new' arguments, the US government simply restated its previous position that market access for foreign semiconductors had not improved sufficiently and much progress still needed to be made. It did though, try to substantiate its argument by saying that without the SCTA there was always the risk of Japanese firms 'back-pedalling', meaning the semiconductor market could gradually close, as old corporate relationships were re-established.[92]

In 1996 there was much political posturing between the US and Japanese governments over the need for a third SCTA.[93] With the foreign market share above the 20 per cent level in 1995,[94] to the Japanese government the need for any form of agreement had long since dissipated. Its position was further strengthened when the figures for the foreign market share for the first quarter of 1996 were released. The figures showed it had soared to 30.6 per cent.[95] The Japanese government's arguments, however, held little sway with the US government. With the foreign market share much less of a concern, the overriding objective of the US government this time around was to consolidate the progress that had already been made. Using its age-old tactics the US government concluded a third agreement with the Japanese government in August 1996. A salient feature of the third SCTA to prevent any 'backsliding' by the Japanese government at the expense of the US semiconductor industry was the establishment of two semiconductor bodies, one multilateral, and the other bilateral. The latter is a private-sector body called the Semiconductor Council, set up by the US government to monitor Japan's semiconductor market. The creation of this body was a US demand that Japan could do little to prevent. One benefit of this body for both sides is that it removed any further direct government intervention from the industry, which the Japanese government wanted to see eliminated from its deregulated market.[96]

Going by the posturing between the US and Japanese governments in the third SCTA, the 'real' purpose of it for the US government, which had long been suspected, was to guarantee US firms a major slice of Japan's valuable semiconductor market.[97] It was no longer about rebuilding the US DRAM industry, which was one of the original objectives of the first SCTA in 1986. Since that date the US DRAM industry has failed to stem its decline. The cycle of decline was reinforced in the late 1990s by the announcements of two major firms that they would leave the sector, Motorola in 1997[98] and Texas Instruments in 1998,[99] leaving Micron as the only US DRAM manufacturer.[100]

Neither was the SCTA about giving protection to US firms who needed time to sort out their production strategies to regain their competitiveness. Macher *et al.* (1998) point out that during the 1990s there was a major change in the strategies of many US semiconductor firms. To reduce their financial exposure many firms went 'fabless'.[101] They specialise in the design of semiconductors, but do not have any production facilities of their own. Interestingly, they rely on foundries, especially Taiwanese ones, for the production of their designs.[102] The 'fabless' semiconductor firm is largely a North American phenomenon, while most of the foundries are located in Asia.[103]

In the late 1990s past experience indicated that the prospects of Japan escaping a fourth SCTA seemed remote. With the bursting of the technology bubble in the early 2000s, the semiconductor industry was struck by an event that was not foreseen and that annihilated at a stroke the US government's chances of imposing another SCTA on Japan. The industry was hit by the worst recession in its history. In 2000 global semiconductor revenues totalled $205 billion. In 2001, however, they fell to $139 billion, a decline of 32 per cent. Collectively, the industry lost $66 billion.[104] The recession was almost twice as bad as the recession of the mid-1980s when the market fell approximately 17 per cent.[105]

The severity of the recession caused enormous turmoil and upheaval. The Japanese semiconductor industry was caught right in the middle. Under major competitive and financial pressures Japanese firms were catapulted into a frenetic process of consolidation and restructuring. A prospect unthinkable a decade earlier, there was an exodus of firms out of DRAMs, which left Japan with only one major manufacturer, Elpida, a joint venture between NEC and Hitachi.[106] By 2002 Japan's global market share in DRAMs had fallen to below 20 per cent.[107] The firms' exit from DRAMs is highly reminiscent of what US firms did following the recession of the mid-1980s. Whilst Japanese firms were reducing their exposure to DRAMs, elsewhere they concentrated their attention on developing alliances with each other and overseas partners to develop more sophisticated semiconductors. Hitachi, for example, integrated its system chip operations with Mitsubishi in 2003;[108] and Fujitsu formed a joint venture with AMD to develop flash memory.[109] The semiconductor recession may have 'humbled' Japan's semiconductor industry but it did not change the US government's appetite for managing the semiconductor industry. The only aspect that changed was its target.

Another firm to fall victim to the savage effects of the industry recession was the South Korean firm, Hynix (formerly known as Hyundai). It was pushed to the brink of bankruptcy, losing $8 billion.[110] Staggering under this weight of debt it would be expected that Hynix would go bankrupt. Much to the incredulity of Hynix's competitors, especially Micron and the US government, Hynix was bailed out with a series of highly controversial loans from its main creditors, mainly state-owned (or controlled) Korean financial institutions.[111] The cause of the controversy was the scale of support given to Hynix, more than $6 billion.[112] Following an abortive merger attempt with Micron,[113] the US Department of Commerce decided to impose tariffs of 44.7 per cent on imported Hynix DRAMs.[114] The Department of Commerce reason for its decision was that Hynix had been unfairly supported by state-backed South Korean banks,[115] an accusation which was strenuously denied by the South Korean government.[116] If by imposing the tariffs the US government had hoped to tip Hynix into bankruptcy the strategy completely failed. Assisted by an improvement in demand for DRAMs and the disposal of non-core assets, Hynix survived the recession and, more incredibly, regained its competitiveness to become the second largest global DRAM maker in 2005.[117] For all the US government's efforts to manage the semiconductor industry, it has never resulted in a US firm seizing industry leadership from either their Japanese or South Koran rivals in DRAMs.

Western Europe

The 'failure' of Western Europe's semiconductor industry to match the competitiveness of the US industry throughout the 1960s to the mid-1970s, or to develop a catch-up strategy akin to the one adopted by Japan, can be explained by firms' 'inertia' in response to important technological changes, the structure and size of demand, and the lack of public policy measures to support the industry while it was still in its embryonic stage. The interaction of these factors led the industry to stall at the very moment it should have been moving into a higher gear. The legacy of this has been that the performance of the European semiconductor industry has consistently lagged behind those of its competitors.

As noted earlier, the US semiconductor industry was very keen to exploit the enormous potential offered by the IC. The dynamism that the US industry displayed was, however, completely lacking in the European industry. Malerba (1985) points out that at the time when the US industry was trying to understand this radically new technology, the European industry was content to remain committed to transistor technology.[118] What led the European industry not to follow the same route as the US industry was the widespread assumption that transistor technology was the 'right' technology to be in.[119] The European industry was dominated by the large vertically integrated receiving tubes producers, namely Philips, Siemens, AEG-Telefunken Thomson-CSF and GEC.[120] To these firms, and others in the industry, the IC was no more than an alternative to the transistor which had a limited number of applications,[121] and germanium technology was more suitable than silicon for small semiconductor devices.[122]

With this blinkered view of the IC and their long familiarity with transistor technology, firms could see no point whatsoever in making the same headlong rush into IC technology as US firms had made, when they could see no immediate or medium-term payoff. Despite the general consensus prevailing within the European industry of the inferiority of IC technology, it did not stop one or two firms from making tentative steps towards developing their own technological IC capabilities. Siemens was one firm to establish R&D and production facilities for ICs in the early 1960s, at the behest of its computer division which thought that one day ICs would become important for use in the production of computers.[123] The firm though did not go ahead with the mass production of ICs until 1968 when it became more economically viable. Philips was another firm that made an attempt to master IC technology, but it was an attempt that was very uncoordinated.[124] For example, in 1965, Mullard, Philips UK subsidiary, produced ICs for a very short time for Marconi computers.[125] Like Siemens, it postponed producing ICs on a commercial basis until later in the decade (1967).[126]

Pressure on European firms to reassess their view of the IC began to mount in the late 1960s when ICs started to be used extensively in a variety of final electronics products.[127] The need for European firms to insure that they had secure and adequate supplies of ICs put pressure on the management of firms to start producing them in-house.[128] But as the importance of ICs gradually began to dawn on European firms, they shied away from embracing the IC 'American style'. The problem for them was that they were in a major fix. To begin with, they were technologically way behind the US firms that dominated the industry. Because of the complexity of IC technology, catching up with US firms was a very tall order. Apart from the resources that they would have had to invest in the technology, it would also have taken them valuable time to become as familiar with the technology as US firms had become. Familiarity with the technology was vital if European firms were to have any chance of achieving the yield levels US firms enjoyed.

On top of that, US firms dominated the European market. Unlike the Japanese government that forced US firms to sell their patents and know-how to establish a presence its market, and barred for many years the establishment of foreign-owned plants, the UK and French governments actively encouraged US firms to set up plants. The overall effect of this was to make indigenous firms even more inclined to retreat into safe niche markets, such as custom-made rather than memory semiconductors (the area where US and Japanese firms competed vigorously). The large presence of the US firms in the European market was more than sufficient to prevent even the 'national champions' from mounting an effective challenge.[129]

European semiconductor firms suffered from an additional disadvantage vis-à-vis the US and Japanese semiconductor firms, the lack of demand from sectors that could have helped the European semiconductor industry to advance at a much faster pace than was the case. As we have seen, the US semiconductor industry benefited hugely from demand for cutting-edge semiconductors, especially from the consumer and military sectors over an extensive period of time. This encouraged the country's semiconductor firms to compete vigorously with each other and put pressure on them to maintain a high level of technological innovation. And the Japanese semiconductor industry enjoyed a high level of demand for cutting-edge semiconductors from its burgeoning consumer electronics industry and later from NTT. Both the US and Japanese semiconductor industries were also able to take advantage of their rapidly expanding computer industries.

One of the key factors which set the European semiconductor industry apart from its US and Japanese counterparts was the level of demand, particularly the demand for specific types of ICs. The largest source of demand for semiconductors was the consumer goods market. It, however, required a different type of semiconductor to, say, that used by the Japanese consumer electronics and computer industries. The Japanese electronics and computer industries used substantial quantities of digital ICs, while the European consumer goods market used more discrete devices and linear ICs. Philips and Siemens, for example, initially specialised in the production of linear ICs as this was the main type of IC which they used in the products that they manufactured.[130] It is worth noting here that there was a very subtle difference between the linear IC and the digital IC: the linear IC was hard to standardise.[131] Linear ICs, therefore, had to be 'custom-made' so they suited the requirements of individual products.[132] For the firms that produced this type of IC, there was one major benefit which most US firms and Japanese firms did not have the luxury of. They were largely sheltered from any real degree of competition and the extreme fluctuations in prices, which were part and parcel of the digital IC market. Therefore, firms found they could survive relatively easily in an otherwise highly competitive industry.[133]

The other major demand factor that severely 'inhibited' the advancement of the European semiconductor industry was the European computer industry. Compared to the US computer market, it was small in size. At no time did the European computer market ever exceed one quarter of the size of the US computer market.[134] The small size of the European computer market did not mean it was an uncompetitive market. European computer firms, if anything, found it too competitive. The underlying cause of European computer firms' problems was the presence of US firms such as DEC, Tandy, Apple, Hewlett-Packard and, of course, IBM, which dominated the Western European market.[135]

In 1967 IBM accounted for 43 per cent of the installed computers in France, 55 per cent in the former West Germany and 29 per cent in the UK.[136] The dominance of IBM made the chances of European computer firms competing with it on a level playing field impossible. Nevertheless, IBM's dominance did not discourage European firms from trying to compete in the standard computer market. Where European firms did try to compete they did not have much luck. Olivetti withdrew from the computer market in 1964, and in the same year, Bull, the largest French computer firm, was acquired by the US firm, General Electric.[137] Philips made a half-baked attempt to enter the computer industry by introducing its first range of computers in 1965, acquiring a computer firm, and taking a substantial stake in another.[138] The crucial element missing from Philips' strategy, which made a nonsense of its attempt to become a player in the computer industry, was a commitment to enter into full-scale production of computers.[139]

With no proper indigenous European computer industry or other forward looking sectors to speak of, the pressures on European semiconductor firms to compete with each other, appreciate important changes in technology, and maintain a high rate of innovation were not present to the same extent as they were in the US semiconductor industry.

The 1970s should have been the decade when the European semi-conductor industry finally came of age. Led by surging demand for computers, consumer electronics, and changes occurring in the telecommunications industry, the demand for ICs soared. Theoreti-cally, as the market for ICs expanded, the European semiconductor industry should simultaneously have entered a period of sustained growth. Unfortunately, the industry could not rid itself of the ghosts that had haunted it in the 1960s. There was no chance of European semiconductor firms growing into a size comparable with US semiconductor firms while they continued to lag behind them technologically, and US firms continued to dominate the European semiconductor and computer markets.

At the very least, any firm that is intent on closing a technological gap with its competitors has to be ready to accept the risks involved and be fully committed to the process from the very beginning. When it came to closing the technological gap with US semiconductor firms, European semiconductor firms showed that they were neither pre-pared to accept the risks, or were in any way committed to the process. Their actions suggested that they simply did not have the stomach for the challenge ahead of them. They only committed them-selves to start producing digital ICs when they recognised that their own long-term future would be in serious jeopardy if they did not. Prior to the mid-1970s, all the major European semiconductor firms whether German, French, British or Italian, continued to operate in 'niche markets'.[140] This type of strategy rapidly became obsolete after the mid-1970s when the demand for ICs underwent a radical and per-manent change. The increasing use of digital ICs in computers, con-sumer electronics, and telecommunications in European and global markets meant that the discrete devices and 'custom-made' semicon-ductors which European semiconductor firms specialised in became less and less important.[141] With their long-term future under threat, the European semiconductor firms committed themselves to the pro-duction of standard digital ICs. It was a process though that was not completed until the early 1980s.[142]

In response to the troubles of the European semiconductor industry, national governments adopted a number of public policy measures aimed at boosting the competitiveness of their own national indus-tries. Prior to 1974, the German government had no policies that were specifically designed to improve the country's performance in digital ICs, primarily because it remained in the dark over the importance of the technology and its potential impact.[143] Before that date the government did provide limited support for the domestic semiconductor industry. Firms like Siemens and AEG-Telefunken received financial support through various government-sponsored programmes, for example, the Data Processing Programmes.

Between 1971–73 DM28.3 million ($8m) was allotted for R&D into circuits, DM16.5 million ($4.7m) for R&D into semiconductors, and DM0.2 million ($57,000) for R&D into ICs.[144] By 1974 the government had become aware of the shortcomings of the electronics industry. It decided the best way to strengthen it was not to target any one technology, but to concentrate on developing the components sector as a whole. In that year the government launched its Electronics Components Programme (1974–78), which focused on five key areas; basic research in new semiconductor and new measurement techniques, DM56.0 million ($21.5m); materials and III–V compounds and materials, DM36.0 million ($13.8m); automated production processes, DM28.5 million ($10.9m); optoelectronics, DM62.0 million ($23.8m); and ICs which included MOS, bipolar, linear, custom and digital, DM85.5 million ($32.9m).[145]

As this programme (and others of a similar nature) focused predominantly on basic and applied research, it did not have the automatic effect that the government had been hoping for, an improvement in the performance of German firms in ICs.[146] As the Japanese semiconductor industry proceeded to expand rapidly, even in the face of US competition, the German government 'discovered' that its Electronics Components Programme had to a certain extent missed the mark. Japan had shown the rest of the world that one of the most efficient ways of enabling the semiconductor industry to grow was to concentrate on production. The success of the Japanese semiconductor industry persuaded the German government it needed to slightly alter its strategy.

In contrast to the Japanese semiconductor firms, which placed great emphasis on product and process innovations, the German government decided it would not attempt to emulate the 'Japanese model'. Instead, it decided the best course of action was to promote the competitiveness of domestic producers in VLSI technology, and support small and medium sized firms in the application of microelectronics.[147] With its first objective, the government devised a special VLSI plan and backed it up with DM125.0 million ($63.6m) of investment between 1979–81.[148] With its second objective, small and medium sized firms received financial help from a number of government sponsored programmes such as the third Data Processing Programme and the Technical Communications Programme, to assist them with microelectronics technology.

The support forthcoming from the government for semiconductors was part of a much wider initiative to improve the general competitiveness of the electronics industry. The pattern of government support remained consistent from the late 1970s through to the mid-1980s. It continued to support a number of different areas within the electronics industry with the launch of its Informationstechnik programme in 1984.[149] This programme focused on the development of semiconductors, other types of electronic components, telecommunications, computers, industrial automation and industrial equipment.[150] At no time does the government appeared to have earmarked the development of ICs for 'special treatment' at the expense of other important technologies.

France's attempt to foster its semiconductor industry has a number of important parallels with that of Germany. One of the major differences between the policies pursued by the German and French governments was that the French government promoted acquisitions among domestic firms and encouraged joint ventures with foreign firms.[151] In parallel with the German government, the French government prior to 1978 had no specific policies to foster an IC industry, but it did give support to its semiconductor firms, namely Thompson-CSF, Sescosem and EFCIS through government sponsored programmes.[152] Financial support came through programmes such as the VI Plan (1971–75) and the VII Plan (1976–81). Through the VI Plan, the government allocated the electronics components sector Fr1,290 million ($235m) and through its successor, the same sector was allotted an additional total of Fr1,850 million ($386m). This was divided up as follows; Fr1,250 million ($250m) for investment subsidies, Fr430 million ($90m) for industrial R&D, and Fr220 million ($46m) for public R&D into active electronic components.[153] Most of these funds went to Sescosem (Thompson-CSF).[154]

Like the German government, in 1978 the French government came to the conclusion that it needed to act when it came to ICs. It was spurred into action by the 1974–75 semiconductor crisis,[155] and the realisation that the IC had become a very important technology. The impact of the IC seems to have had a much more powerful effect on it than on the German government. The French government devised its very own Integrated Circuit Plan (1978–82), which included a 'minimum' of Fr600 million ($133m) for R&D and investment subsidies.[156] The objective was to build an internationally competitive IC industry, an ambitious plan under any circumstances. It was all the more so considering that French firms had virtually no presence in the technology when the plan was drafted.[157] The plan was structured so that certain firms were 'made' to specialise in specific types of ICs. The government encouraged firms to go along with the plan with the use of subsidies. For example, Thompson-CSF was given Fr100 million ($22.2m) over five years to produce bipolar ICs for consumer electronics; Eurotechnique-Complec (RTC) (then a subsidiary of Philips) was likewise given Fr100 million ($22.2m) to produce high-speed bipolar ICs for the computer and telecommunications markets; and EFCIS received Fr200 million ($44.6m) to specialise in MOS technology, especially NMOS, SOS and CMOS.[158]

The government might well have perceived its Integrated Circuit Plan as a bold piece of thinking on its behalf. However, when it came down to the practicalities the government soon realised that the gap between theory and reality was enormous. Broadly speaking, the government lacked a thorough knowledge of the semiconductor industry and how it functioned, so when it came to implementing the plan, the government ran into trouble. The plan was criticised on a number of fronts. Firstly, it took a year and a half for the plan to be completed, which is far too long for a rapidly moving industry.[159] Secondly, the level of financial support was far too low for the government to expect firms to close the technological gap with US firms; and lastly, the financial support was spread too thinly among too many firms which made their task of achieving the vital economies of scale more difficult than it might otherwise have been.[160]

The initial criticisms of the plan appear to have been warranted, as it subsequently proved ineffective. In 1983 the socialist government that had been elected in 1981, launched a second integrated circuit plan.[161] This plan was part of its Filiere Electronique programme, which was meant to turn the domestic electronics industry into a major global player. One of the most important aspects of the second plan compared to the first, was the level of investment which the government was to provide to the semiconductor industry. Under the second plan the government increased its level of investment more than six-fold in dollar terms. Between 1983–87 the semiconductor industry was to receive a total of Fr6 billion ($820m), Fr3.4 billion ($465m) for R&D, and Fr2.2 billion ($300m) for investment subsidies.[162]

Another major aspect of the second plan was the way in which the government allocated funds among firms. In the first plan, it had been criticised for spreading funds too thinly among too many firms.

Anxious to avoid making the same mistake twice, this time round the government ensured funding went to just two firms, Thompson-CSF and Mantra. Sescosem, EFCIS and Eurotechnique-Complec, which had previously received government funding under the first plan, did not lose out under this second plan. The government made absolutely sure of that by simply merging all three with Thompson-CSF, which it nationalised along with Mantra.[163] After merging with three other firms, Thomson-CSF became France's largest IC producer.[164] It might have become big, but the actions of the French government did not transform it into a global player.

Similarly in the UK, the turning point for government policy with respect to semiconductors was 1978. Government policy changed significantly from a regime of 'minimal' support to one of direct intervention. It was a course of action that the UK government assumed would lead to the creation of a competitive IC industry.

Throughout most of the 1970s, government policy was aimed at creating a competitive computer industry, which forced semiconductors to take a back seat. The great bulk of financial support with which the UK government provided the semiconductor industry came through its two main programmes, the Microelectronics Support Scheme and the Computer Industry Scheme (1976–79).[165] In the first programme, the government invested £12 million, distributed among GEC, Plessey and Ferranti, to support R&D in microelectronics that would otherwise have not been carried out.[166] In the second programme, the government invested a total of £20 million in electronic components, covering some twenty-two different groups. Out of the £20 million, £10.8 million was to support R&D, and £7.4 million was for investment subsidies. Out of the twenty-two different groups of electronic components, semiconductors received the highest level of support of approximately £5 million.[167] Besides these two programmes, another important source of funding for the main semiconductor firms, GEC, Plessey, Ferranti, STC and Mullard, was the Ministry of Defence (MOD). Between 1971–76 the MOD awarded more than 75 per cent of its contracts to these firms alone.[168]

As in France, the main catalyst that helped bring about the major sea change in UK government thinking was the 1974–75 semiconductor crisis. When the UK government discovered that there was a conspicuous absence of UK firms in digital ICs, in particular, it came to the same conclusion as the French government did when it discovered that French firms were largely absent in this key technology; it was time for action. In the years following 1978, UK government policy for the semiconductor industry was much more comprehensive than it had ever been before.

As an integral part of its policy for a vibrant IC industry, the government decided that one of the quickest ways of developing a major presence in ICs was to set up a firm that specialised in them. So in 1978, it set up a publicly owned firm known as INMOS which specialised in digital ICs (initially 16K DRAMs) and microprocessors.[169] INMOS received £65 million from the government in cash and £35 million in loan guarantees as of May 1984.[170] Another major plank of the government's strategy for ICs was a raft of government-sponsored programmes; two of the largest of these were the Microelectronics Support Programme and the Microprocessor Applications Programme.[171] The Microelectronics Support Programme was in many ways just another version of an earlier programme, the Computer Industry Scheme. The main differences between the two were the amount of investment and the sector being supported. They provided £70 million, which was later cut to £55 million by the Conservative government, for R&D support and investment subsidies to semiconductor firms located in the UK.[172] Like similar programmes run by the German and French governments, the UK programme covered 25 per cent of productive investments made by firms, and between 25–33 per cent of their R&D expenditures.[173] In addition, the Microprocessor Applications Programme was of a similar nature to the German programme that provided financial support to small and medium sized firms to help them with microelectronics technology. Under the programme, £55 million was invested by the UK government to promote the diffusion of microelectronics applications throughout the industry. In 1982 the programme was extended when the government allotted it an additional £30 million.[174]

A feature of UK government policy from the late 1970s, which distinguished it from the semiconductor policies of the German and French governments, was that it had no criteria which determined what firms could qualify for government support and those that should be excluded. The French government, for example, was only willing to give support to a foreign firm so long as it was linked up with a French firm, perhaps through a joint venture, and the foreign firm's own interests did not run contrary to those of the French state. In retrospect, the UK government had little option but to pursue a 'liberal' policy of support if the UK was to rapidly build up a domestic IC industry. With indigenous UK firms extremely weak in ICs, the next best policy option was to lure firms irrespective of their nationality to invest in the UK. The major beneficiary of this liberal policy were US firms, for example,

National Semiconductor received in excess of £5 million when it decided to build a production facility in Scotland to produce ICs, and ITT received £2 million to produce ICs in the UK.[175]

In the 1980s, UK government policy continued very much along the lines that had been established in the late 1970s. Foreign semiconductor firms continued to be lured to invest in the UK, and the government continued to support the development of the technological capabilities of the domestic semiconductor industry. As far as the latter was concerned, the two major programmes which the government hoped would improve the competitiveness of UK firms were the Microelectronics Industry Support Programme II (1984–90) and the Alvey Programme (1983–88). The first programme was allotted a total of £120 million, and focused primarily on supporting the development of custom and semicustom ICs and their use.[176] The second programme focused on four key areas, but special prominence was given to supporting long-term basic research in VLSI technology for which £100 million was made available, and computer aided design (CAD) for which £30 million was made available.

The intervention of the German, French and UK governments in their national semiconductor industries during the 1970s and early 1980s, did not, however, give the European semiconductor industry the competitive edge it desperately needed. Any hopes that the European semiconductor industry may have had of some breathing space with the waning of US dominance quickly evaporated with the emergence of Japan.

The weakness of European semiconductor firms has continued into the 1990s and beyond. In the mid-1990s, the situation as it stood was that firms from Japan and South Korea rather than those from the US, supplied 80 per cent of the $5.76 billion market,[177] but European firms only had 1 per cent of the Japanese market, 6 per cent of the US market,[178] and 9 per cent of the Asian market (excluding Japan).[179] Due to plunging prices in early 1997, the EU was prompted to impose anti-dumping duties on semiconductors made by 11 Japanese and South Korean firms to help alleviate some of the competitive pressures European firms have been under. The 80 per cent drop in the price of a 16M DRAM to around $9 had once again exposed the weaknesses of European semiconductor firms. In almost every instance, whenever conditions have got particularly uncomfortable in an industry well known for extreme swings in supply and demand, they have sought some form of protection. When there has been a major downturn in the market, European semiconductor firms have normally clamoured for the imposition of anti-dumping measures.

Maybe in recognition that these are normally only of a temporary nature, and have not proved a long-term solution to the European semiconductor firms' competitive problems, the EU has chosen to adopt a similar strategy to that used by the US government and 'manage' the industry. At the end of 1997, the EU concluded an agreement modelled on the SCTA with South Korea. It is an industry-to-industry agreement that allowed for the anti-dumping tariffs (which had been imposed in 1993 and 1997) on South Korean firms to be lifted, and provided a mechanism for swift investigations into any future dumping complaints when there was another severe downturn in the industry.[180] The agreement contained no minimum pricing element, but it required firms to collect data on production costs and export prices.[181] The mechanism for swift investigation into dumping complaints was put into effect when the recession in the semiconductor industry struck in 2001 and took its toll on European semiconductor firms, namely Infineon. Following a complaint from the firm, the EU launched its own investigation into Hynix.[182] Following in the footsteps of the US, it imposed tariffs of 33 per cent on imported Hynix DRAMs in July 2003, citing the same argument used by the US government.[183] As with US experience, the EU has tried to 'manage' the semiconductor industry with the explicit objective of improving the competitiveness of European semiconductor firms and failed. To add to the European industry's troubles, despite a successful WTO ruling in mid-2005 which declared the tariffs imposed were legal, it is still no nearer to finding a solution to its Achilles heel: the question of its long-term competitiveness. In the past this has proved an intractable problem and in the future it threatens to amplify itself even more unless action is taken soon.

Summary

From the analysis here, we can see that the interrelationships between the factors at work in the semiconductor industry are complex to say the least. For example, the effectiveness of public policy is intimately tied up with important supply and demand factors and the technological choices of firms at critical junctures. An obvious factor which links the growth of the US and Japanese industries and stunted the growth of the European one is the role played by the user, which in this case was the computer industry. The main factors present in the semiconductor industry that have had a major impact, and continue to do so are public policy, very rapid technological change, wild fluctuations in supply and demand, intense competition, firms' behaviour, falling prices, and the role of the user. These same factors are also evident in the TFT-LCD industry and will become more evident in the following chapters. However, the effects which each of these factors have had on the semiconductor industry vary, sometimes quite considerably, from those which they have had on the LCD industry, for example, in relation to public policy which has played only a relatively minor role in the development of the LCD industry. Unlike public policy's minor role, fluctuations in supply and demand, intense competition, firms' behaviour, and the role of the user, have all played a major role in the LCD industry's development.

Notes

1. Braun, E & MacDonald, S (1980) *Revolution in Miniature: The History and Impact of Semiconductor Electronics*, Cambridge: Cambridge University Press, p. 67.
2. *Ibid.*
3. *Ibid*:54.
4. *Ibid*:78.
5. *Ibid.*
6. Tyson, L (1992) *Who's Bashing Whom: Trade Conflict in High Technology Industries*, Washington, DC: International Institute for Economics.
7. Braun & MacDonald, 1980:79.
8. Tyson, 1992:86.
9. Braun & MacDonald, 1980:79.
10. *Ibid*:80–81.
11. *Ibid.*
12. *Ibid*:80.
13. *Ibid.*
14. *Ibid*:88. Figures obtained from Table 7.3.
15. *Ibid*:102.
16. *Ibid*:105.
17. *Ibid*:115.
18. *Ibid.*
19. Borrus, M, Millstein, JE & Zysman, J (1983) 'Trade and Development in the Semiconductor Industry: Japanese Challenge and American Industry' in L Tyson & J Zysman (eds) *American Industry in International Competition: Government Policies and Corporate Strategies*, New York: Cornell University Press, p. 152. This same figure is also quoted by Tyson, 1992:90.
20. Wilson, RW, Ashton, PK, & Egan, TP (1980) *Innovation, Competition and Government Policy in the Semiconductor Industry*, Massachusetts & Toronto: Lexington Books, p. 153.
21. Uenohara, M, Sugano, T, Linvill, JC & Weinstein, FB, in DI Okimoto, T Sugano, & FB Weinstein (eds) (1984) *Competitive Edge: The Semicon-*

ductor Industry in the US and Japan, Stanford, California: Stanford University Press, p. 22.
22. Borrus *et al.*, 1983:174.
23. Figures obtained from Tyson, 1992:104, Table 4.3.
24. Forester, T (1993) *Silicon Samurai: How Japan Conquered the World's IT Industry*, Oxford: Blackwell, p. 65.
25. Weinstein, FB, Uenohara, M, & Linvill, JC, in Okimoto *et al.* (eds), 1984:53.
26. *Ibid.* Forester, 1993:65.
27. *Ibid.*
28. Tyson, 1992:101. Weinstein *et al.*, in Okimoto *et al.* (eds), 1984:54.
29. Weinstein *et al.*, in Okimoto *et al.* (eds) 1984:53.
30. Fransman, M (1990) *The Market and Beyond: Cooperation and Competition in Information Technology in the Japanese System*, Cambridge: Cambridge University Press, p. 24.
31. *Ibid*:95.
32. Weinstein *et al.*, in Okimoto *et al.* (eds) 1984:55.
33. Tyson, 1992:93.
34. For an in-depth analysis of the VLSI programme see Fransman (1990) Chapter 3.
35. Forester, 1993:65.
36. *Ibid.*
37. This figure has been calculated from Tyson, 1992:104, Table 4.3.
38. *Ibid*:105.
39. *Ibid*:103.
40. Forester, 1993:73.
41. Tyson, 1992:101.
42. Forester, 1993:62.
43. Malerba, F (1985) *The Semiconductor Business: The Economics of Rapid Growth and Decline*, London: Pinter, p. 152.
44. Tyson, 1992:105.
45. Malerba, 1985:155.
46. Macher, JT, Mowery, DC & Hodges, DA (1998) 'Reversal of Fortune?: The Recovery of the US Semiconductor Industry', *California Management Review*, Fall 1998, Vol. 41, No. 1, p. 111.
47. Weinstein *et al.*, in Okimoto *et al.*, 1984:35.
48. Tyson, 1992:97.
49. *Ibid*:106.
50. Macher *et al.*, 1998:124.
51. Tyson, 1992:109.
52. *Ibid*:110.
53. *Ibid.*
54. *Ibid.* Mowery, DC & Rosenberg, N (1989) 'New Developments in US Technology Policy: Implications for Competitiveness and International Trade Policy', *California Management Review*, Vol. 32, No. 1, p. 114.
55. Tyson, 1992:110.
56. *Ibid.*
57. *Ibid.*
58. *Ibid.*

59. *Ibid*:111.
60. *Ibid*:112.
61. *Ibid*.
62. *Ibid*.
63. *Ibid*.
64. *Ibid*.
65. *Ibid*.
66. *Ibid*:114.
67. *Ibid*.
68. *Ibid*:111.
69. *Ibid*:115, Table 4.4.
70. *Ibid*.
71. *Ibid*.
72. *Ibid*.
73. *Ibid*:113.
74. *Ibid*:114.
75. *Ibid*. The effect of the SCTA on Japanese prices was to push them up only very marginally.
76. *Ibid*:117.
77. The two firms accounted for 55 per cent of Japan's 256K DRAM exports.
78. Tyson, 1992:117.
79. *Ibid*.
80. *Ibid*:131.
81. *Ibid*.
82. *Ibid*.
83. *Ibid*.
84. 'Japan "trapped" by chips import deal', *Financial Times*, 23 March 1994.
85. 'Tokyo "backsliding" on chips accord – Trade relations in jeopardy, US manufacturers warn Clinton', *Financial Times*, 9 February 1994.
86. 'Japan "trapped" by chips import deal', *Financial Times*, 23 March 1994.
87. 'Jump in foreign chip sales', *Financial Times*, 19 March 1994.
88. 'Foreign chips take 21.9% of Japan's market', *Financial Times*, 15 September 1994.
89. *Ibid*.
90. 'Washington and Tokyo split on renewal of the US-Japan semiconductor agreement – Kantor calls for market share pact to be reviewed. No need, say the Japanese, it has already worked', *Financial Times*, 23 November 1995. In 1994 the foreign market share was approximately 23 per cent.
91. *Ibid*.
92. *Ibid*.
93. 'Japan braced for pressure on US access', *Financial Times*, 5 February 1996.
94. 'Japan chip market more open', *Financial Times*, 18 June 1996. In the first quarter of 1995, the foreign market share was 22.8 per cent. By the last quarter of 1995 the figure had increased to 29.8 per cent.
95. 'Japan's chip market more open', *Financial Times*, 18 June 1996.
96. 'US and Japan in agreement on chips', *Financial Times*, 3 August 1996.

97. 'US and Japan close to deal on microchips', *Financial Times*, 2 August 1996.
98. Survey of Semiconductors: 'Risk factors for producers: A perilous business', *Financial Times*, 4 February 1998.
99. 'Texas Instruments quits memory chips', *Financial Times*, 19 June 1998.
100. *Ibid*.
101. Macher *et al*., 1998:119.
102. *Ibid*. For an exposition of how Taiwan developed its semiconductor industry see Mathews, JA & Cho, DS (2000) *Tiger Technology: The Creation of a Semiconductor Industry in East Asia*, Cambridge: Cambridge University Press, Chapter 4.
103. Macher *et al*., 1998:119.
104. 'The chips are down, but hope still flickers: In spite of the recent travails of the world semiconductor industry, innovation is flourishing as chip suppliers invest in wireless data and networking technologies', Survey Section, *Financial Times*, 17 April 2002.
105. *Ibid*.
106. 'Chipmakers find benefits in cooperation', *Financial Times*, 26 September 2002.
107. *Ibid*.
108. *Ibid*.
109. 'Fujitsu, AMD in flash-memory talks', *Financial Times*, 9 October 2002.
110. 'Hynix condemns US move to impose tariffs', *Financial Times*, 19 June 2003.
111. 'South Korea maintains foreign focus: Resistance to a Hynix/Micron deal has opened old wounds', *Financial Times*, 24 April 2002; 'Pressure builds on Seoul over Hynix; Creditors are contemplating a third multi-billion dollar bailout of the troubled chipmaker amid mounting protest', *Financial Times*, 9 December 2002 (USA Edition).
112. 'Hynix condemns US move to impose tariffs', *Financial Times*, 19 June 2003. The form of support has been a mixture of loans and debt relief.
113. 'Micron ends hopes of Hynix merger', *Financial Times*, 3 May 2002.
114. 'Hynix condemns US move to impose tariffs', *Financial Times*, 19 June 2003.
115. *Ibid*.
116. 'South Korea denies Hynix subsidy claims', *Financial Times*, 23 November 2002.
117. 'Hynix fails to hit profit forecast on probe provision', *Financial Times*, 4 March 2005.
118. Malerba, 1985:105.
119. *Ibid*.
120. *Ibid*:119.
121. *Ibid*:121.
122. *Ibid*:105.
123. *Ibid*:107.
124. *Ibid*.
125. *Ibid*:106–107.
126. *Ibid*:107.
127. *Ibid*:110.

128.   *Ibid*.
129.   *Ibid*:108–119.
130.   *Ibid*:123–124.
131.   *Ibid*:123.
132.   *Ibid*.
133.   *Ibid*.
134.   *Ibid*:125.
135.   *Ibid*:175.
136.   *Ibid*: Figures obtained from Table 5.23.
137.   *Ibid*:127.
138.   *Ibid*.
139.   *Ibid*.
140.   *Ibid*:156–162.
141.   *Ibid*:162.
142.   *Ibid*.
143.   *Ibid*:190.
144.   *Ibid*.
145.   *Ibid*:191.
146.   *Ibid*.
147.   *Ibid*:192.
148.   *Ibid*.
149.   *Ibid*.
150.   *Ibid*.
151.   *Ibid*:193.
152.   *Ibid*.
153.   *Ibid*.
154.   *Ibid*.
155.   *Ibid*.
156.   *Ibid*.
157.   *Ibid*:194.
158.   *Ibid*.
159.   *Ibid*.
160.   *Ibid*:194–195.
161.   *Ibid*:195.
162.   *Ibid*.
163.   *Ibid*.
164.   *Ibid*.
165.   *Ibid*:196.
166.   *Ibid*.
167.   *Ibid*.
168.   *Ibid*.
169.   *Ibid*:197.
170.   *Ibid*.
171.   *Ibid*.
172.   *Ibid*.
173.   *Ibid*.
174.   *Ibid*.
175.   *Ibid*.
176.   *Ibid*:198.

177. 'Chips are down as EU acts on dumping', *Financial Times*, 1 April 1997.
178. 'Brussels steps up chip pact pressure', *Financial Times*, 31 July 1996.
179. 'EU protest over chip pact', *Financial Times*, 12 March 1996.
180. 'Anti-dumping duties lifted on Korean D-Rams', *Financial Times*, 1 December 1997.
181. *Ibid.*
182. 'EU probes South Korean chipmakers', *Financial Times*, 26 July 2002.
183. 'Hynix to prevail in WTO ruling over EU says report', *Electronic Engineering Times*, 16 March 2005 (www.eetimes.com).

5
Corporate Strategy and Firm Learning

Corporate strategy is a complex subject with many different facets. Put differently, it is a field where consensus is rare and disagreements widespread. In the first chapter of his book, *What is Strategy – And Does it Matter?*, Whittington (2001) shines the spotlight on the core problem and how it goes to the very heart of the subject:

> There is not much agreement about strategy. *The Economist* (1993:106) observes: 'the consultants and theorists jostling to advise businesses cannot even agree on the most basic of all questions: what precisely, is a corporate strategy'. Strategy guru Michael Porter (1996) asks the question 'What is Strategy?' in the very title of an important Harvard Business Review article. In a recent textbook, Markides (2000:vii) admits: 'We simply do not know what a good strategy is or how to develop a good one'.[1]

This is some statement to make. Whittington's motivation for doing so would seem to arise from the fact that views on the role of corporate strategy have oscillated wildly and this appears to be a continuing trend. *The Economist* article quoted by Whittington casts considerable light on the extent of this change. In the immediate aftermath of the second world war corporate strategy had strong military overtones. The prevailing management philosophy was to 'attack' markets and 'defeat' rivals. This idea rapidly fell from grace with the realisation there were few parallels between killing the enemy and outselling them.[2] In the 1960s corporate strategy became equated with excessively detailed corporate planning based on forecasts of economies and specific markets. This view was given legitimacy with the publication of two well known books, *My Years with*

General Motors by Alfred Sloan, the man who built GM into the world's largest industrial firm, and *Strategy and Structure* by Alfred Chandler, who documented the history of large multidivisional successful US firms. It was the rise of competition from Japan in the 1970s which shattered the idea that detailed corporate planning was the way forward. Successful Japanese firms had no time for it. The next view of corporate strategy to follow the corporate planning perspective was Michael Porter (1980) with his book *Competitive Strategy*.[3] He argued that a firm's profitability was determined by the characteristics of its industry and the firm's position within it, so these should also determine its strategy.[4] Coming from a background in industrial economics, Porter's premise was that the goal of the firm should be to find a position in the industry where it could best defend itself from competitors, either through product differentiation or by becoming a low-cost producer. This way it would be able to generate higher profit margins or erect entry barriers to deter new entrants.[5] Although *Competitive Strategy* was a successful book, the ideas contained in it did not prove as influential as may be imagined. Porter's downfall was the nature of his ideas which did not tell firms what they should and should not do.[6] At the beginning of the 1990s Gary Hamel & CK Prahalad formulated their core competence perspective and argued that the role of strategy should be to 'stretch' the firm, management should find out what the limitations of its capabilities are. More than a decade later, Hamel & Prahalad's perspective remains influential, but it has not quelled the enduring debate to which Whittington refers: What is the role of corporate strategy?

This is very much in evidence in the texts to be found. Johnson & Scholes (2002) and Lynch (2003) agree that the subject is generally concerned with an organisation's long-term future, however, the overall emphasis is quite different between the former and the latter.[7] For the former, the primary focus of corporate strategy is about an organisation establishing a competitive advantage within its environment, whilst for the latter it is more about an organisation's purpose and how it interacts with the environment. A more appropriate summation comes from Kay (1993) who contends that corporate strategy is more concerned with matching markets to a firm's distinctive capabilities.[8] Put this way, much of the 'vagueness' of the subject is eliminated. In his analysis, Kay shows success is based on *added-value* which takes time to achieve and must be hard won. His stance is at odds with much of the literature in that he takes a much more pragmatic approach to the topic. This fundamental inconsistency is again put into an illuminating perspective by Whittington:

> Amazon.com lists forty-seven books available with the title Strategic Management. Most are thick tomes, filled with charts, lists, and nostrums, promising the reader the fundamentals of corporate strategy. Cursory inspection reveals that they nearly all contain the same matrices, the same authorities. There is little variety, little self-doubt. These texts generally sell for around $50.
>
> There is a basic implausibility about these books. If the secrets of corporate strategy could be acquired for $50, then we would not pay our top managers so much. If there really was so much agreement on the fundamentals of corporate strategy, then strategic decisions would not be so hard to make.[9]

There may not be much general agreement about what corporate strategy is about nor how to develop a successful one, but Lynch (2003) asks the next logical question: What makes 'good' strategy? The answer he gives focuses on three core elements; first, the need to bring added-value to the market place, either in the form of increased profitability, market share, or enhanced innovative ability; second, a strong element of consistency and flexibility; and finally, the strategy will be a form of sustained advantage to the firm.[10] The literature is very limited on how firms should develop a strategy akin to this. In the analytical framework used by Kay to understand the foundations of firm success, he identifies the capabilities that the firm needs to develop a good strategy: architecture (which is its set of contractual relationships), innovation, reputation, and strategic assets. Each firm he finds is unique in relation to these capabilities and he uses the examples of BMW and Honda as evidence of the long-term successful management and exploitation of these capabilities. The purpose of this chapter then is not to provide a comprehensive overview of the corporate literature *per se*, but to bring a new dimension to it – the level of importance that firms attach to learning. Thus the chapter is structured to examine two contrasting mainstream strategies used by firms to gain a competitive advantage. The emphasis is on the *degree* of learning each strategy engenders and how it affects the firm's capacity to learn. The selection of one strategy over the other provides a 'reliable' guide to how strong the commitment to learning is.

The competitive environment and the firm

There are two areas in corporate strategy where there is a definitive consensus. The first concerns the importance of the firm's 'environment'. According to Mintzberg (1993) the environment 'comprises virtually everything outside the organisation – its "technology" (the knowledge base it must draw upon); the nature of its products, customers, and competitors; its geographical setting; the economic, political, and even meteorological climate in which it must operate and so on'.[11] The value of Mintzberg's definition is that it achieves three important objectives; it tells us what the term environment means; it identifies the key elements in the firm's environment; and illustrates the environment's diversity. The second area is the instability of the environment. The firm's environment is persistently unstable and the only factor which differentiates one environment from the next in this context is the rate of change. Change directly challenges the flexibility of the firm, so the issue for the firm becomes how it adapts its strategy and responds to change.

With change comes uncertainty and this is something the firm has no option but to confront. In many respects uncertainty is a double-edged sword in so far as it poses the firm with threats and opportunities. As uncertainty is such a major issue it has generated much debate between strategists over how it can be managed and what its principal sources are. Disagreements are bound to occur as some factors will be more important than others depending on the sector. Furthermore, it is more than possible for the same factor to affect two firms in the same sector in very different ways. Dependent upon the situation, the firm can react in one of two ways in the environment. It can be proactive, that is, it can develop a strategy to exploit an opportunity, or alternatively, it can be reactive, forced to adapt to a situation over which it has no control. Ideally, a firm should be proactive, but to be proactive the firm needs to carefully analyse its own competitive environment, or more specifically, its own industry.

There can be little doubt that one of the most influential frameworks used to carry out such an analysis to date has been the work of Porter (1980) and his 'five forces' framework. In this framework the industry is the main unit of analysis, where it is assumed a group of firms produce goods which are of a similar nature to one another. The sources of competition are: (i) the threat of new entrants; (ii) the bargaining power of buyers; (iii) the bargaining power of suppliers; (iv) the threat of substitute products and services; (v) rivalry amongst established firms. Porter goes to considerable lengths to explain how each of these sources act to stimulate competition.[12] The firm can cope with the five forces by following one of three potentially successful generic strategies; overall cost leadership, differentiation, and focus. The only other alternative route open to the firm is to get 'stuck in the middle', which is 'an extremely poor strategic situation'.[13] This is disastrous for the firm as it is hit by low profitability. It will therefore lack the requisite resources to follow one of the three generic strategies to restore its fortunes. Porter concedes that the firm can get out of this unsatisfactory position through depending on its own resources and capabilities and being totally focused on the generic strategy it aims to pursue. Whatever strategy the firm chooses it must be mindful it will take time and sustained effort to turn the corner.[14]

Porter's framework alerted the firm to the importance to gain a more in-depth understanding of the dynamics of its own environment. The very structure of the framework, however, has exposed it to considerable criticism as industry analysis is made to look 'black and white'. The framework has been criticised on a number of different fronts; Lynch points out that the framework is essentially static, when the reality is that the competitive environment is continuously changing; and Johnson and Scholes comment that the five forces are not independent of each other, the inter-relationships are much more complex in the mainstay of sectors than the framework assumes. Whittington (2001) comments that the framework ignores the role of government and labour; Pitkethly (2002) states that no allowance is made for creativity; whilst Tidd *et al.* (2005) highlight the facts that it underestimates the power of technological change to transform industrial structures and says nothing about the problems of implementing strategy.[15] It is this last point of Tidd *et al.*, the problem of implementation, where firms have a tendency to come unstuck, which is the focus of the next section. Successful implementation rests on the firm having the requisite internal skills and resources or acquiring them externally. In the following chapters, understanding why firms opt for a specific strategy is key to 'unlocking' their success or lack of it in semiconductors and TFT-LCDs.

Strategy and implementation

When implementing a strategy, management may focus on the specific factors which it considers are the most appropriate, whilst at the same time ignoring others at its peril. As in Porter's five forces framework, management needs to understand that factors are not independent of each other. Much of the success of management hinges upon the generation of synergy. The implementation process is problematic at the best of times. The firm always has to navigate its way around numerous obstacles. Yet it is this search for synergy that is probably the most hazardous of them all.

The magnitude of the challenge facing management can be deduced from Sirower (1997). He shows that what (top) management all too frequently attempts to do is risky and the probability of failure is high. The high level of risk is built into the 'formula', 2+2=5, which according to Sirower, 'is the common definition of synergy'.[16] It is reasonable to ask why firms still show a preponderance to follow this treacherous path? The answer is that synergy is all about competing better. Sirower offers his own definition of synergy *'as increases in competitiveness and resulting in cash flows beyond what two companies are expected to accomplish independently'*.[17] From a managerial perspective, the fundamental problem with this version of synergy is the firm is expected to gain a competitive advantage well beyond what it would otherwise require to survive in the market place.[18]

The main area where management looks for synergy is through mergers and acquisitions (M&A). Booms in this area can occur because the pressure on management to grow revenues and profits year on year is intense. Bandwagons develop because firms are all in pursuit of the same objective: the creation of a portfolio of businesses which is worth more than its businesses would be worth as stand alone entities. This is the typical method used by firms to secure the synergy they all want so badly. Management's infatuation with mergers and acquisitions is solely to do with the fact that it's the quickest way to increase the value of a business. Mergers and acquisitions can be completed within weeks or months, whereas the development of new products takes years.[19] As the payoff is potentially very considerable in terms of both time and profitability, management continues to harbour the strong belief that so long as it can manage mergers and acquisitions it can secure the synergy it believes is within its reach. However, what a high percentage of management would prefer to ignore is that the probability of failure is high.

One of the factors which contributes to the failure of mergers and acquisitions is that the acquiring firm pays between 25 per cent and 50 per cent above the pre-bid share price for the acquisition. Davidson (2002) likens the buying of a firm to a cross between buying a house and attending an auction. Management gets emotionally involved and competitive instincts become inflamed and reason goes out of the window.[20] The premium paid represents the alleged benefits of the acquisition.[21] Davidson puts the failure rate anywhere between 50 per cent and 70 per cent. The figures may appear excessively high, but in comparison to the figures for new product development the rate is low. The failure rate for new product development is between 75 per cent and 90 per cent.[22]

Another cause of failure is the overestimation of the benefits of a merger or acquisition.[23] The management of the acquiring firm will make the best possible case for the acquisition, frequently in strict financial terms such as the projected cost savings and revenue gains. Invariably, these figures look very attractive from a distance. The integration element of the merger or acquisition is totally overlooked. This should not come as any great surprise as this is precisely the area where problems can originate. The challenge that management faces is to demonstrate how the merger or acquisition will make it harder for the acquiring firm's competitors to compete in the market place before it has made its first move. To have an outside chance of success the acquiring firm ideally needs to open up new markets/and or encroach on the competitors' markets where these competitors cannot respond.[24] By keeping attention focused away from integration and on the financial fundamentals, this makes the task of enhancing the attractiveness of the merger or acquisition a good deal easier.

Once a bandwagon has started to gain momentum there can be no telling how far it will roll or what the eventual consequences will be. The 1990s were a decade which experienced one such bandwagon. On one day alone, 22 April 1996, with the announcement of the Bell Atlantic-Nynex merger and Cisco Systems acquisition of Stratacom, over $27 billion of acquisitions were announced. For 1995, the total value of acquisitions activity was over $400 billion.[25] In 1999 the amount spent on mergers and acquisitions totalled $3.3 trillion, up 32 per cent from the previous year and equivalent to 35 per cent of US GDP.[26]

When management embarks on a merger or acquisition it has to contend with some in-built flaws in the system that can lead to major problems. For example, when firm A decides it wants to acquire firm B for its strong presence in a technology (or assets) where A is weak, firm A is willing to pay a premium over and above the current market value of firm B. For firm A to take control of firm B it has to pay up front and that is just to 'touch the wheel'. Here is where the problems for the acquiring firm can begin. The first pitfall for management, where it risks being 'blinded', is during the purchase process. As the system demands the acquiring firm must pay up front there is a good chance of purchasing a firm about which it does not have all the necessary information. It is akin to a customer who purchases a vehicle off a garage forecourt but has no idea of its roadworthiness. The customer can only find this out once the vehicle has been paid for and it can then be driven on the road. In the meanwhile, all the customer can do is hope the vehicle is mechanically sound, whilst appreciating its exterior. This inability of the acquiring firm to fully examine the acquisition prior to the purchase puts it in a quandary. In order to take control of the acquisition it has to proceed with the purchase; on the other hand, proceeding with the purchase can leave it in possession of an acquisition it later discovers it cannot fully utilise for all sorts of reasons.

A second pitfall is created by the payment of the premium, which automatically puts the firm under additional pressure from the equities market. The equities market already expects the firm to meet its targets, but the premium 'signals' to the equities market that the firm's position will be improved, putting it in the position to meet higher targets. The equities market's expectation is that this improvement will rapidly feed through into better performance. Quite frequently, a firm cannot move as fast as the equities market expects and fails to meet market expectations. Falling foul of the equities market can see it lose the premium and more. Sirower (1997) neatly encapsulates the precarious position the acquiring firm puts itself in:

> To visualise what synergy is and what exactly the premium represents in performance terms, imagine being on a treadmill. Suppose you are running at 3 mph but are required to run at 4 mph next year and 5 mph the year after. Synergy would mean running harder than this expectation with competitors supplying a head wind. Paying a premium for synergy – that is the right to run harder is like putting on a heavy pack. Meanwhile the more you delay running harder, the higher the incline is set. *This is the acquisition game.*[27]

With its eye firmly fixed on the equities market, the firm can sideline other aspects of the merger or acquisition, which actually need an equal amount of attention. Integration is one such aspect where there is evidence that this occurs. Sirower cites a major study by the Boston Consulting Group that found that during the pre-merger stage, eight out of ten firms did not even consider how the acquired company would be integrated at the operational level after the acquisition.[28] This kind of managerial short-sightedness is another explanation why a firm can fall victim to the equities market. Improvements in performance cannot be expected to materialise when management does not get around to attending to these matters until much later in the day than it should. To take one example, the integration of information systems is always a complex operation and cannot be done overnight. The merging of systems can frequently be hampered by issues related to compatibility and glitches in software which can lead to delays running into weeks and months. Sirower goes as far to say that integration can (in some instances) take years. When management analyses integration the factor it has most to fear from is time. The acquiring firm has signalled to its competitors that it expects to become more competitive. It is under pressure to complete the process in the quickest possible time. The longer the process takes the greater the likelihood the firm's rivals will respond before the firm can make any noticeable improvements in its competitiveness. Management never wants to factor these 'unforeseen' problems into the equation. Superficially, what can seem a good fit on paper can in practice be analogous to fitting a square into a circle – the two don't go together.

A salutary example of a firm which grossly underestimated the integration process is AT&T with its merger with NCR. In 1991, AT&T authorised a payment of $4.2 billion to acquire NCR, a premium of 125 per cent above the pre-bid share price, having already lost approximately $2 billion in its own computer business between 1985 and 1990.[29] For reasons unknown, AT&T left the NCR executives in their positions for two years following the acquisition. Unusually, they were asked to 'find' synergy for AT&T's old computer and marketing business in the expectation that computer and communications technologies would converge. The problem for AT&T was that its strengths were in telecommunication switches and not in computers. The synergies did not materialise and in 1993 revenues came under pressure. In what can only be seen as a knee-jerk reaction to a difficult situation, AT&T initiated the process to implement its strategy for NCR. Most of the senior management at NCR left the firm between 1993 and 1995 and AT&T set up a whole new set of sales teams in over 100 different countries, pushing AT&T into new markets where it had no experience. The eventual fall out was that in 1995 AT&T had lost the entire premium it had paid for NCR and suffered additional losses of $720 million for that year alone.[30]

There are other more recent examples of mergers and acquisitions which have gone wrong. Hewlett Packard's acquisition of Compaq is one such case. In 2002 Hewlett Packard spent $19 billion acquiring Compaq to create a major new force in the computer industry. The strategy of Hewlett Packard was based on the assumption that its increase in size would allow it to restore the profitability of its personal computer business and realise economies of scope to challenge IBM in the corporate market.[31] Central to this strategy was a major cost-cutting programme. Ironically, the programme was well executed.[32] The widely anticipated synergy, that is, the increased profitability of its personal computer business had not materialised by early 2005. What caught Hewlett Packard out and severely undermined its strategy was the commoditisation of the personal computer market and the nimbleness and improved competitiveness of Dell, its major rival.[33] By focusing its strategy on cost-cutting Hewlett Packard found itself on a treadmill where it had to run very hard to stand still. Hewlett Packard is now in a situation where it must completely revise its entire strategy before it is in any position to decide which direction it should go in next.

An even more high profile merger that has not worked out as originally planned is Daimler-Benz and Chrysler. The merging of these two household names would on paper appear to make sense. Needless to say, ever since the firms merged in 1998, DaimlerChrysler has been hit by a series of major problems. For a start, it has seen the group's market value collapse. From a post-merger high of €94 billion in April 1999, the group's value had declined to €35 billion in 2005, which was *less* than what Daimler was worth prior to the merger.[34] As discussed earlier, integration is commonly neglected when two firms merge and DaimlerChrysler is no exception to this rule. Following the merger, Daimler faced an enormous task to get Chrysler back into profit. It has managed to do this at a cost of two financial crises and the elimination of 36,000 US jobs. Quality has been another source for concern, especially for Daimler. It has been badly hit by the highly complex electronic systems used in its automobiles. This has necessitated the recall of 1.3 million cars, nearly one in three of the 4 million cars it has sold in the past four years.[35] The electronic systems have proved so taxing because Daimler found it did not possess all the necessary engineering expertise to fully understand how they functioned. Daimler has for decades been synonymous with quality, however, this well deserved reputation has been greatly tarnished with these high profile problems. The quality problems have not received the management attention they deserved and as a consequence Daimler-Benz has fallen from first place in the influential American JD Power vehicle reliability survey in the past decade to twenty-eighth place out of 37.[36] DaimlerChrysler now faces a long haul to get back on track.

Given the above examples, it is not hard to see why Hamel & Prahalad (1994) argue that when a firm chooses to go down the mergers and acquisitions route in the vain search for synergy, it is doing little more than chasing a mirage. Their supposition is that the way firms can secure the synergy they yearn for is to strike off in an entirely new direction. Their basic tenet is that where management has long gone wrong is with its preoccupation with the present and lack of attention on the future. This preoccupation with the present stems from the inordinate amount of time management spends on it. They cite figures that show on average management spends less than 3 per cent of its time on building a corporate perspective on the future. This figure drops to less than 1 per cent in some firms.[37] For far too long management has spent its time focusing on short-term issues such as 'downsizing' and 'reengineering'. These so-called strategies have their rightful place at the appropriate time, however, they are no substitute for preparing the firm for the demands of tomorrow. The grave risk for management is that unless it refocuses the firm's strategy on the markets of the future it will find any upturn in its fortunes will be short-lived. After all, no firm can prosper in new industries and markets if it lacks the foundations on which it can build.

In the recession of the early 1990s firms embarked on a large-scale programme of downsizing in the face of the combined pressures of the economic downturn and the increasing intensity of competition. For example, in 1993, large US firms announced nearly 600,000 layoffs, 25 per cent more than in 1992, and nearly 10 per cent above the levels of 1991, which was technically the bottom of the recession in the US.[38] As the economic cycle turned up, firms wanted to start expanding again and they looked for the fastest way to grow after the lean years they had experienced. Their automatic reaction was to look in the direction of mergers and acquisitions, which laid the foundations for the eventual boom.

Management's penchant for mergers and acquisitions signifies something more fundamental to Hamel & Prahalad – how management has become subordinate to the needs of the market. In essence, they see mergers and acquisitions as no more than management opting out of its responsibilities to confront the future:

Top management often sees a major acquisition as the only escape route from a business that has become hopelessly mature. It's not news to anyone that few acquisitions actually benefit the sharehold-ers of the acquiring company, yet acquisitions are, in many cases, an easy way out for senior executives too intellectually lazy to think through the future of the firm's 'core business' and too unimaginative to discover new ways of deploying existing capabilities.[39]

The reluctance of management to make more provision for the future is understandable to a point. Management has a thorough knowledge of the present, which is not something that can be said about the future. The message that Hamel & Prahalad want to convey to management is it can no longer afford to close its eyes to the future. Preparing for the future demands a lot from management. The poten-tial rewards are immense for the firm prepared to make the move and put in the effort. By ignoring the future management also needs to be made aware that there is an additional price to be paid by the firm in the form of 'lost' synergy. This will remain beyond management's grasp until such time as it decides to act. The type of synergy that Hamel & Prahalad advocate relates to firms' core competencies. The obvious advantage of this type of synergy is that it can give a firm a major long-term and sustainable competitive advantage. As pointed out in Chapter 1, Western firms have rarely seen their core competen-cies as a source of competitive advantage. Japanese firms, and more recently South Korean and Taiwanese firms, see them as a very potent and effective competitive weapon. What differentiates the two types of synergy is that the latter involves an extensive period of firm learning, which the former does not.

In Chapter 1, it was mentioned that Sony is held up as an enduring example of a firm whose core competencies in miniaturisation have long been the source of its innovative products. The outstanding success that Sony has traditionally enjoyed is testimony to the rewards this type of synergy can generate for a firm when managed properly. In the early 2000s it had lost its competitive edge in key product seg-ments in consumer electronics having been usurped by the likes of Apple's iPod and flat screen televisions produced by Sharp and Samsung. This is the heart of management's problem. This type of synergy is difficult to manage, it requires significant investment and time to develop. The firm has to be very watchful on this front because if it takes its eye off the ball, the cost can be the loss of industry leader-ship. To reiterate the point, core competencies need to be protected and continually renewed. Hamel & Prahalad note that by the end of the 1980s a large number of firms which had begun the decade with their industry leadership intact had lost it. These included some major household names: IBM, Philips, Xerox, Boeing and Daimler-Benz. The conundrum that management must resolve to achieve this preferred synergy is to strike the correct balance between the demands of the present and those of the future.

The starting point is management's own role in the way it views and develops strategy. Management must be aware that it is about competing to shape the structure of the industries of tomorrow and how it can do this.[40] This requires an analysis of the changing nature of competition and the convergence of industries. This invaluable exercise can provide management with useful insights into the emerging patterns of the future and the comparisons that can be drawn with the present. The nature of competition has been radically altered by a range of different driving forces: globalisation, the rapid pace of technological change, shortening product life-cycles and increased consumer expectations. These driving forces have put management under considerable pressure to be much more creative and 'visionary' and to anticipate how the industries of the future will evolve before their competitors can make any sort of move. The need to act should not be underestimated. Being a first-mover in a growing market or a 'fast follower' will affect a firm's ability to secure a stream of future profits, develop alliances, and build up a consumer base. It is in the best interests of the firm to gain as big a head start over its rivals as possible as industry leadership in fundamentally new industries can take anywhere between ten and fifteen years.[41] Furthermore, if the firm develops its core competencies in conjunction with the evolution of a new industry, it can control the destiny of the firm and shape the industry at the same time.

For some, understanding the competition would seem easy, whilst for others this is less so. Intel is one firm that has long known how to keep any potential competitors at bay and reinforce its dominance by maintaining a high rate of technological innovation and pushing forward the frontiers of microprocessor technology. Contrast this with the experience of General Motors (GM), which for four decades has continued to lose market share in its home market, principally to Japanese automobile firms. Its failure to become as efficient and flexible as its Japanese competitors can be attributed to its lack of understanding and its late implementation of 'lean production'. This system of production, pioneered by Toyota, dates back to the 1950s and the firm has been continuously refining it ever since, but it took GM more than 40 years to understand its significance.[42] Hamel & Prahalad note 'Toyota's foresight had become GM's implementation nightmare'.[43]

This hard lesson, that GM has learned to its enormous cost, exemplifies how the ground can shift in very subtle ways from incumbent firms, going virtually unnoticed until it is almost too late. This form of unorthodox competition can prove extremely difficult to counteract if the basis of it is not well understood. As GM evidences, when the nature of competition changes the maturity of the industry makes it no easier to deal with, irrespective of the firm's size and resources. Competition can be changed by domestic firms or overseas competitors, either way the challenge still has to be met head on. The examples of Intel and GM deal with competition where it already exists, but there is another element to competition from which firms can reap huge rewards; identifying new markets and demand where they do not yet exist.

The learning organisation

For a firm to create a new market and demand for a product (or service) where none existed before requires it to combine vision and innovation from the start. To use Senge's (1990) own terminology the firm has to become a 'learning organisation' to move forward. His hypothesis is that the majority of firms suffer from a learning disability, an inherently strong belief that they can learn through their own direct experience. He casts doubt on this by pointing out that firms inevitably establish organisational routines and patterns of behaviour that become deeply embedded with the course of time. From this perspective the ability of firms to learn is impeded by their 'learning horizon'. Senge likens the firm to an individual with his notion of the learning horizon. In the case of the individual, the learning horizon is 'a breadth of vision in time and space within which we assess our effectiveness when our actions have consequences beyond our learning horizon, it becomes impossible to learn from direct experience'.[44] This is the main dilemma for the firm in the context of learning. They never have direct experience of the consequences of their own actions. The repercussions of their most important decisions may not be apparent for years or even decades.[45] Two of the areas where key decisions can have profound effects are R&D and the promotion of key personnel to positions in top management. With the former, this can 'ripple' through into the functions of marketing and manufacturing, and with the latter, new management shape strategy and can create an organisational culture that can prove very enduring. To emphasise his point about how restrictive a firm's learning horizon can be, we need to return to our previous example, the US automobile industry.

Often firms cannot see how and where their problems originate because of chronic management myopia. Hamel & Prahalad offer an alternative perspective to Senge's and postulate that firms also need to learn to forget. They show how firms can exhibit some of the same characteristics as the dinosaur. Failure to adapt will threaten the long-term survival of the firm. Although unlike the dinosaur, which became extinct through its distinct inability to adapt hindered by its 'genetic coding',[46] the firm does not suffer the disadvantage of a fixed genetic profile. Previously, it was stated that GM took 40 years to understand the significance of lean production. Between the 1950s and late 1980s the US market share of the 'Big Three' automobile manufacturers continued gradually to decline. How did they fall into this trap? The Japanese automakers were not seen as a serious competitive threat in 1962 when their market share was less than 4 per cent, in 1967 when it was less than 10 per cent, in 1974 when it was less than 15 per cent, and in the early 1980s when it was 20.3 per cent.[47] The automobile industry is not the only industry to have suffered from this limited learning horizon. Other American manufacturers have proved equally susceptible to it, when they worked on the assumption that there was a trade-off between cost and quality. Competition from Japan showed them no such correlation existed.

Management myopia permeates the firm through the establishment of a dominant management philosophy. It is also known as the Hayes & Abernathy syndrome. Evidence of its continued longevity can be found in Whittington (2001). Interestingly, his research reveals the key factors that help guide the making of strategy at the international level and highlights the very marked differences between various countries. Whittington cites the valuable work of Hitt *et al.* (1997) who carried out a comparative analysis of US and South Korean managers' strategic decision making using 15 criteria.[48] For US managers, the three most important criteria were (in order of magnitude); projected demand, projected cash-flow, and return on investment (ROI). This compared with the South Korean managers whose three most important criteria were (in order of magnitude) industry attractiveness, sales growth, and market share.[49] The Japanese approach to the use of financial considerations is very similar to the South Korean's; it is low down on their list of major priorities. Its main use appears to be more as a guiding principle to inform on strategy than as a basis on which strategic decisions are made.[50] Like the South Koreans, the Japanese continue to give higher priority to aggressive growth or market domination than to considerations of short-term profitability.[51]

Recall that ROI is one of the major characteristics of the Hayes & Abernathy syndrome. Whittington reveals how it has become so deep rooted in US firms. He traces its origins back to the 1920s when newly diversified firms such as Du Pont and GM developed and used financial performance indicators as tools to analyse the performance of their different businesses in their portfolios. The quarterly financial report emerged as the standard measurement of the internal efficiency of the firm (or division) and the management of business 'by numbers' became an accepted practice. With the rise of the new management orthodoxy, the use of short-term financial considerations were used more and more to determine strategy.

Every firm faces the same issue of resource allocation. Decisions on how and where a firm invests its resources will lead to some areas of the firm being strengthened, and others being left alone, enabling the firm to pursue some opportunities and not others. The pattern of the firm's investment will affect its overall competitiveness over the short and long term. The firm's environment is constantly changing and this continually provides the firm with opportunities for investment. How and when a firm takes advantage of these opportunities is dependent upon its management.

Management's most obvious task is to make the most out of each opportunity it seeks to exploit. This is only possible if its investment decisions are made in a timely and appropriate fashion. Investment decisions are, however, dependent on what Hayes *et al.* (1988) call the managerial infrastructure.[52] This is the firm's resource allocation system. Management can draw up very carefully laid out plans and make a carefully considered decision to exploit a specific opportunity which entails the firm going off in a particular direction. However, there is always the risk that the best laid plans can go wrong if the management infrastructure 'malfunctions'. Hayes *et al.* state that perhaps the most difficult of all investment decisions to make are those based on a proposal to invest current resources in the prospect of future returns.[53] The system used by firms to get around this problem is the capital budgeting system. Once established, it will affect the firm's competencies and overall level of competitiveness. Various sophisticated financial tools and techniques have been developed, such as discounted cash-flow analysis to aid firms with their investment making process and to assess the likely impact on the firm. The crucial aspect about the capital budgeting system is that it serves a dual purpose: it differentiates and selects between alternative projects and influences how the firm defines and comes to view projects.[54]

How a firm views projects, especially those which are of a risky nature will by definition be determined by how far the Hayes & Abernathy syndrome has managed to permeate its management system. In the cases where it has permeated the entire system the projects that will become most attractive to the firm will be those where the element of risk is minimal and the payback period short. The payback period is based on the concept of the discounted present value. The basic concept of the idea is that a dollar is worth more today than at some point in the future.[55] How much more depends on the other investment opportunities available. Taken too literally, the firm can be lured into thinking that if it can select out projects where it can safeguard its initial investment and generate an 'acceptable' level of profits it will be home and dry. The firm, however, has to be very watchful with this approach as it has a major drawback. The problem is the payback period. It is seen from a static standpoint and takes no account of the changes to the firm's environment during the actual payback period, or as Hayes *et al.* point out, the period following the end of the payback period.[56] It all starts with how projects are evaluated. Projects are normally evaluated using a set of established criteria. The criteria used should provide the firm with an assessment of the likely advantages and benefits of a project. No allowance is made for anything other than the continuation of the *status quo*. The purpose is to provide the firm with an objective and accurate assessment of a scenario of the potential ramifications should the project not be given the go-ahead.[57]

One of the criteria used is the return on investment (ROI). Hayes *et al.* contend that the ROI, or 'hurdle rate', that firms use in the evaluation process is set too high. The range can be anywhere between 25–40 per cent. These rates are well above the average cost of capital. Two of the explanations used by firms to justify these rates are that they protect the firm from the uncertainties of the investment process (such as a sudden drop in revenues) and act as a tool to motivate management. The theory is that if tough targets are set management will put in the extra effort to hit them, as they are used as a performance indicator to measure an individual manager or management groups' effectiveness. Under these conditions a good project would be one with a well executed discount flow analysis and the appropriate ROI. The firm can be taken in by thinking that important investments can be postponed at no cost. This though does not take account of the wider picture, in other words, what its competitors are doing. A firm's position can start to deteriorate in an industry and arresting or reversing the situation can prove extremely difficult if not impossible. As profitability declines and the situation deteriorates for the firm, the investment required to reverse the process mounts rapidly. The rising cost of investment exacerbates an already difficult situation for the firm and pushes it into a vicious cycle where it has no other choice than divestment and to leave the industry altogether. As Hayes *et al.* observe, stress on the capital budgeting paradigm encourages the firm to pay far too much attention to the investment project itself than to the wider implications of that investment.[58] Furthermore, it stymies its capacity to learn about new technologies and markets.[59] When a firm has been permeated by the Hayes & Abernathy syndrome, the idea that it has the potential of becoming a 'learning organisation' is a complete non-starter.

Approaches to strategy

The above analysis has drawn out the key advantages and disadvantages associated with two major strategies adopted by firms to gain a competitive advantage. To recap briefly, the main advantage of mergers and acquisitions for the firm is it *can* be the quickest way to increase the market value of a firm. The obvious disadvantage is that the likelihood of failure is high. With core competencies, the key advantage is that they are a long term and sustainable source of competitive advantage for the firm when they are hard to replicate. The major disadvantages are the time and resources required to develop them. The area that needs further elaboration is why these strategies were chosen. One engenders an extensive period of learning by the firm and the other does not. When a firm chooses one strategy over the other, it provides a good indication of their commitment to learning. A firm that opts for a mergers and acquisitions strategy can be categorised as having a low commitment to learning. Alternatively, a firm that opts for a core competence strategy can be categorised as having a high commitment to learning. Through this it can be explained why firms from some countries are better at learning about technology that others. From the evidence here, it is suggested that mergers and acquisitions are mainly a Western phenomenon and core competencies an East Asian phenomenon. A mergers and acquisitions strategy encourages a short-term focus. A firm requires a long-term strategy (and focus) to learn about technology successfully, especially in dynamic industries. As Hitt *et al.* (1997)

show countries have their own distinct approaches to strategy. These differences have major implications at the level of the firm. The approach that South Korea has towards strategy allows its firms to adopt a long-term strategy towards technology. The same can be said for Japanese and Taiwanese firms. The opposite is true for US and Western European firms. Success, especially in high-technology industries depends on solid foundations. A long-term strategy provides for this, whereas a short-term strategy does not.

Summary

The major conclusion to be drawn from this chapter is that the development of a good strategy, that is, a successful strategy, poses huge challenges to the firm. Although the firm faces a considerable uphill struggle, the adoption of a long-term strategy with an associated high level commitment to learning will significantly increase the probability of successful future development. It needs to be borne in mind that the development of such a strategy is firm-specific. Unless a firm has a degree of commitment to learning its chances of success are much diminished.

Notes

1. Whittington, R (2001) *What is Strategy – And Does it Matter?'*, London: International Thomson Learning Business Press, 2nd edition, p. 2.
2. The sources Whittington refers to are: Markides, C (2000) *Making All the Right Moves: A Guide to Crafting Breakthrough Strategy*, Boston, Massachusetts: Harvard Business School; Porter, M (1996) 'What is Strategy', *Harvard Business Review*, November–December, pp. 61–78; 'Eenie, meenie, minie, mo...', *The Economist*, 20 March 1993.
3. Porter, M (1980) *Competitive Strategy: Techniques for Analysing Industries and Competitors*, New York: The Free Press.
4. 'Eenie, meenie, minie, mo...', *The Economist*, 20 March 1993, p. 106.
5. *Ibid.*
6. *Ibid.*
7. Lynch, R (2003) *Corporate Strategy*, Financial Times, Harlow: Prentice Hall, 3rd edition; Johnson, G & Scholes, K (2002) *Exploring Corporate Strategy*, Financial Times, Harlow: Prentice Hall, 6th edition.
8. Kay, J (1993) *The Foundations of Corporate Success: How Business Strategies Add Value*, Oxford: Oxford University Press, p. 17.
9. Whittington, 2001:1.
10. Lynch, 2003:22.
11. Mintzberg, H (1993) *Structure in Fives: Designing Effective Organisations*, New Jersey: Prentice Hall, p. 136.
12. For recent interpretations of Porter's five-forces framework see Thompson, JL (2003) *Strategic Management*, London: Thomson Learning, pp. 290–300;

Tidd, J, Bessant, J & Pavitt, K (2005) *Managing Innovation: Integrating Technological, Market and Organisational Change*, 3rd edition, Chichester: John Wiley & Sons, pp. 121–125; Johnson & Scholes, 2002:112–120; Pitkethly, R (2002) 'Analysing the Environment', in *The Oxford Handbook of Strategy*, *Strategy Overview and Competitive Strategy*, Oxford University Press, Vol. 1, pp. 225–260.
13. Porter, 1980:41.
14. *Ibid*:42.
15. Tidd *et al.*, 2005:124.
16. Sirower, M (1997) *The Synergy Trap: How Companies Lose the Acquisition Game*, New York: The Free Press.
17. *Ibid*:6. Author's emphasis.
18. *Ibid*:20.
19. Davidson, JH (2002) *The Committed Enterprise: How to Make Vision and Values Work*, p. 268.
20. *Ibid*.
21. *Ibid*.
22. *Ibid*: footnote 6.
23. *Ibid*:270.
24. Sirower, 1997:25.
25. *Ibid*:6.
26. Davidson, 2002:266.
27. Sirower, 1997:10. Author's emphasis.
28. *Ibid*:7.
29. Sirower, 1997:34–5.
30. *Ibid*.
31. 'Firing Ms Fiorina: Her successor should consider breaking up the company', *Financial Times*, 10 February 2005.
32. *Ibid*.
33. *Ibid*.
34. 'DaimlerChrysler: Jurgen Schrempp Interview', *Financial Times*, 4 March 2005.
35. 'Electronic bugs cause recall of 1.3 million cars by Mercedes', *Financial Times*, 1 April 2005.
36. 'DaimlerChrysler: Jurgen Schrempp Interview', *Financial Times*, 4 March 2005.
37. Hamel, G & Prahalad, CK (1994) *Competing for the Future*, Boston, Massachusetts: Harvard University Press, p. 4.
38. *Ibid*:7.
39. *Ibid*:85.
40. *Ibid*:25.
41. *Ibid*:37.
42. *Ibid*:13.
43. *Ibid*:82.
44. Senge, P (1990) *The Fifth Discipline: The Art and Practice of the Learning Environment*, London: Century Business.
45. *Ibid*:22.
46. This is the term used by Hamel & Prahalad.
47. Senge, 1990:22.

48. Hitt, MA, Dacin, M, Tyler, B & Park, D (1997) 'Understanding the Differences in Korean and US Executives' Strategic Orientations', *Strategic Management Journal*, Vol. 18:2, 159–167.
49. *Ibid*:164.
50. Whittington, 2001:64.
51. *Ibid*:96.
52. Hayes, RH, Wheelwright, SC & Clark, KB (1988) *Dynamic Manufacturing: Creating the Learning Organisation*, New York: The Free Press, p. 61.
53. *Ibid*.
54. *Ibid*:64.
55. *Ibid*:64.
56. *Ibid*:63.
57. *Ibid*:74.
58. *Ibid*:81.
59. *Ibid*:84.

6
Liquid Crystal Displays as an Emerging Sectoral System of Innovation

The first section examines the historical background of the LCD. In Chapter 2, the notion of path dependency was explained at some length. The assertion made was that to analyse any 'unusual outcome' with this approach, it was imperative to find out where the path begins. For understanding the complexities of the development of the LCD industry, the path begins here. It was the pioneering work of two US firms, as stated in Chapter 1, which paved the way for the commercial exploitation of liquid crystals, and gave the US its lead in this technology. The second section focuses on why these highly innovative firms subsequently allowed LCD technology to slip through their fingers, and gave away the country's lead to Japan.

The origins of LCDs

The origins of the industry date from around 1850 and 1888 when groups of scientists in Europe, working on three different types of experiments detected highly unusual forms of chemical behaviour never witnessed before.[1] The experiments focused on the study of biological specimens, how substances crystallise, and compounds synthesised from cholesterol. In the experiments on biological specimens, it was noticed through a microscope, that the outer cover of a nerve fibre formed soft and flowing forms when left in water, and furthermore, these forms produced unusual effects when exposed to polarised light. Normally polarised light affects solids, but not liquids. The biological specimens, however, were not solids and reacted in an unusual way when subjected to polarised light. Collings (1990) notes that the importance of this reaction was not immediately appreciated as biological structures are complicated and knowledge about them was far from extensive.[2]

In the second set of experiments on how substances crystallise, it was a German physicist, Otto Lehmann, who made one of the earliest and most fundamental contributions to the field of liquid crystals. He developed a very important piece of laboratory equipment that was to revolutionise research into liquid crystals. He made a heating-stage that could be used in conjunction with a polarising microscope. According to Collings, this is now a standard piece of apparatus for any laboratory that carries out research into liquid crystals.[3]

With his heating-stage, Lehmann was able to observe in a controlled environment how materials crystallised at different temperatures. Prior to the development of the heating-stage this type of experiment had been impossible. What Lehmann observed was that some substances did not crystallise from a clear liquid, but changed instead to an amorphous form, which then crystallised.[4] Often it is the presence of impurities that adversely affect chemical reactions, but on this occasion this was not the problem. In fact, what he had observed was a completely new type of phase transition. Lehmann though, did not know this when he made this observation. What he was very well aware of was that phase transitions do not follow any specific pattern. They vary considerably depending on temperature, any impurities present, and the substances involved.[5]

In the final set of experiments on compounds synthesised from cholesterol, it was discovered that they produced striking colours when cooled. In addition, it was observed that the compounds produced a very unusual chemical reaction – they did not automatically change from a normal liquid into a solid or vice-versa. Little did the scientists know they had come across a phase of matter which was distinct from a solid or a liquid.[6]

The value of all these experiments only becomes clear when they are looked at in relation to how liquid crystals were 'discovered'. The man who has been widely attributed with discovering liquid crystals in 1888 was an Austrian botanist, Freidrich Reinitzer, whose main area of interest was the functioning of cholesterol in plants. He noticed the organic chemical, cholesteryl benzoate, had two different melting points, 145.5°C and 178.5°C. At the former temperature it became a cloudy liquid and at the latter temperature it turned into a clear liquid.[7] His results were remarkably similar to the results of the experiments on compounds synthesised from cholesterol, and another experiment mentioned very briefly by Collings (1990).[8]

Reinitzer was aware of the work of Lehmann and recognised that there was a connection between his own work and Lehmann's work.[9] In an attempt to establish exactly what he had come across, Reinitzer sent Lehmann some samples to get his opinion on them.[10] He wrote to Lehmann:

> Encouraged by Dr V. Zepharovich (Prof of Minerology at Vienna), I venture to ask you to investigate somewhat closer the physical isomerism of the two enclosed substances. Both substances show such striking and beautiful phenomena that I can hopefully expect that they will also interest you to a high degree…The substance has two melting points, if it can be expressed in such a manner. At 145.5°C it melts to a cloudy, but fully liquid melt which at 178.5°C suddenly becomes completely clear. On cooling a violet and blue colour phenomenon appears, which then quickly disappears leaving the substance cloudy but still liquid. On further cooling the violet and blue coloration appears again and immediately afterwards the substance solidifies to a white crystalline mass. The cloudiness on cooling is caused by the star shaped aggregate. On melting of the solid the cloudiness is caused not by crystals but by a liquid which forms *oily streaks* in the melt.[11]

The results of Lehmann's subsequent analysis showed Reinitzer's findings had remarkable similarities with Lehmann's own findings, and with those who had worked on the experiments mentioned above. When Lehmann investigated the cloudy intermediate state he discovered that although cholesteryl benzoate was a viscous liquid, the chemical possessed many of the properties of a crystalline solid, so he named the combination of these properties 'liquid crystals'.[12] Although Lehmann can be credited with coming up with an appropriate name that encapsulates these unusual chemical properties; it was Reinitzer's original work on cholesterol substances that helped pioneer a field which has been intensely researched over the course of the last century.[13]

From the late 1880s until the turn of the century when research activity reached a peak, interest in liquid crystals grew considerably and knowledge about them advanced quite rapidly. After this period of intense activity had passed, widespread interest in liquid crystals did not completely evaporate as Heilmeier (1976) has claimed, but it did wane quite substantially.[14] During this period research on liquid crystals certainly continued in France, Holland and Germany. Largely because of the work of Lehmann, Germany developed a centre of excellence at the Martin Luther University in Halle, which has carried out research into liquid crystals since 1900.[15] What happened during the inter-war period seems to be a matter of some dispute. Heilmeier claims there was only a brief period between the late 1920s and early 1930s which saw a resurgence of interest,[16] but Kelker (1973) and Collings (1990) show that from 1922 through to the beginning of the war, pioneering work was done in a number of countries. For example, theoretical work on the elastic properties of liquid crystals was done in Sweden and England, which later helped explain why liquid crystals adopted various orientational configurations. Work in France and Germany revealed that liquid crystals possess more order than liquids, but less than solids.[17] Following the end of the war, there was yet another period of inactivity which lasted until the late 1950s, when there was a renewed interest, particularly in cholesteric liquid crystals. In retrospect, this was to prove the real turning point for liquid crystals as research has continued to flourish since then, albeit in different countries.

The main reason for the resurgence of interest in liquid crystals in the late 1950s appears to have been the interest of specific individuals keen to advance the understanding of the molecular structure, optical properties, and technical opportunities.[18] In 1962 the English chemist, George Gray, who later made a fundamental contribution to the industry in the early 1970s, published an important book on the *Molecular Structure and Properties of Liquid Crystals*.[19] In Germany two physicists, Wilhelm Maier and Alfred Saupe, developed a major new theory that predicted the behaviour of the liquid crystal phase. The theory claimed that the behaviour of individual molecules could be predicted as it quickly became apparent that molecules were highly sensitive to changes in temperature, electromagnetic radiation, and the chemical environment.[20] In themselves, these were significant contributions to the understanding of liquid crystals, but it was research done in the US by two major firms, Westinghouse and RCA, that was to revolutionise the whole industry and created the opportunity for the commercial exploitation of LCDs over the coming decades.

The early US experience

During the 1950s Westinghouse and RCA were highly innovative firms, spending large amounts of capital, stimulated by the growth of the semiconductor industry. For Westinghouse profitability was its major priority, whilst for RCA, remaining at the forefront of technology was

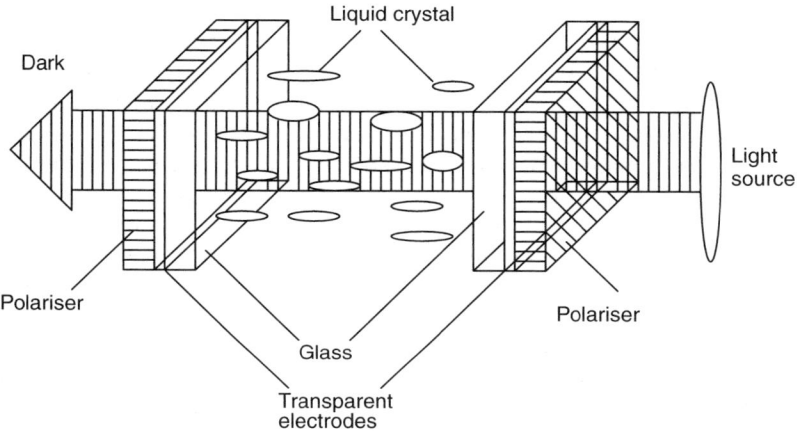

Dark

Liquid crystal

Light source

Polariser

Glass

Transparent electrodes

Polariser

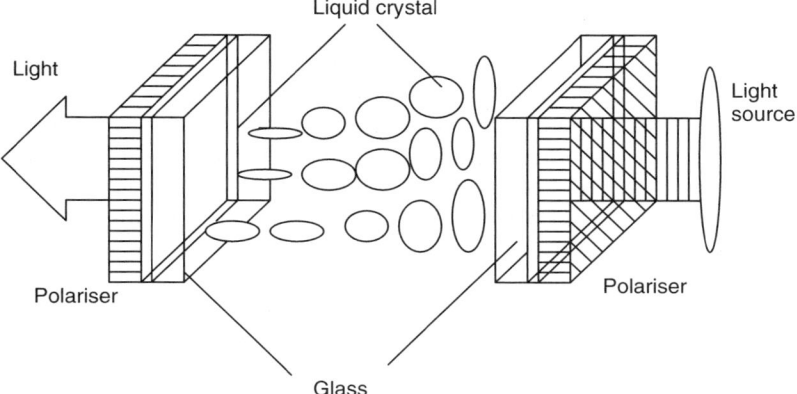

Light

Liquid crystal

Light source

Polariser

Polariser

Glass

Figure 6.1   The basic construction of a twisted-nematic LCD
Source: O'Mara, W (1992) 'LCD dividends', *Physics World*, Vol. 5, No. 6, June, p. 38

its key priority, especially as it prided itself on being one of the country's most technologically progressive firms. The issue of short-term profitability was never really much of a consideration at RCA unlike at Westinghouse. The main reason being that its main area of specialisation was developing technologies that it could license, which obviously is something which cannot be achieved overnight. The revenue RCA generated from the technologies it licensed to other firms were then ploughed back into the firm's extensive research activities.

Despite their very different philosophies, both firms ended up abandoning the very technologies that they had pioneered. Initially, this seems a very odd outcome, but on closer inspection it should come as no surprise. Gradually, both Westinghouse and RCA became afflicted with at least one of the main elements of the Hayes and Abernathy syndrome.

Between the late 1950s and late 1960s Westinghouse made two fundamental contributions to the LCD industry. Its first was in 1958 when it discovered a method which showed how liquid crystals could be used as a display technology on a commercial basis. Its second contribution was the development of the core technology that is now used in the third generation of LCDs, thin-film transistors (TFTs), which is discussed below. What can be considered to be the ultimate breakthrough was made by RCA in 1963 at its David Sarnoff Research Centre. It discovered that when liquid crystals are subject to an electric charge, the molecules could be realigned so they either passed or blocked polarised light.[21] This discovery, and Westinghouse's earlier work on the commercial potential of liquid crystals, effectively laid the necessary foundations for the commercial exploitation of the first generation of LCDs, known as the twisted-nematic (TN). The two major applications of the TN-LCD were in digital watches and calculators throughout the 1970s. The limited usage of the TN-LCD was largely due to its construction, which by modern standards is very simple. It comprised of two glass plates pressed together with the liquid crystal located between the plates. When the liquid crystal is subject to a charge the liquid crystal molecules 'twist' to an angle of 90°, so the light which passes through the polariser on the right is able to produce an image on the left. When not subjected to a charge the liquid crystal molecules are aligned at 0°, that is, an 'untwisted' state (see Figure 6.1).

RCA

The discovery that RCA made put the firm in an unrivalled position vis-à-vis its rivals, whereby it had an excellent window of opportunity to develop a highly promising new technology that could have been a valuable source of future expansion. According to Heilmeier who was central to the firm's research into LCDs, in 1964 the idea of, 'The wall sized flat panel colour TV was just around the corner – all you had to do was ask us!'.[22] In fact, he was head of an interdisciplinary team that built the world's first prototype LCD in 1968.[23] For most firms, such a lead in a promising new technology is often an opportunity too good to be missed, but surprisingly, RCA let this opportunity slip through its fingers.

RCA appear to have let this opportunity slip away as a result of conflict between the management of RCA Laboratories who needed no convincing of the value of liquid crystals and the management of RCA's product divisions.[24] From Heilmeier's account of events at the time, the strong impression given is that the future of liquid crystal research and development was rendered bleak by the attitude adopted by the management of the product divisions. As Heilmeier states:

> Liquid crystals were viewed more as a threat than any opportunity. After all, 'it wasn't silicon', the materials were 'dirty' by semiconductor standards and it was 'too easy for the garage operators' to get into the business. When customers began clamouring for more information attempts were made to discourage them or to point out that what they really wanted was a Numatron display.[25]

With the major decision making effectively in the hands of the management of the product divisions, they were the ones who decided which projects were rejected and approved. As they were heavily biased against liquid crystals, feeble excuses were made for the rejection of two key ideas; the concept of the digital watch, and the possibility that a liquid crystal cell could be filled and sealed between two pieces of glass (one with an evaporated film on the surface and the other with a tin or indium oxide film on the surface).[26] The former was dismissed on the grounds that there was 'no market', and the latter because it was 'insurmountable'.[27] This is very odd considering that in 1971 RCA had shown to watchmakers, a watch which the firm's Solid State division had developed that displayed the time using liquid crystal digits.[28] The management's outright refusal to pursue these ideas led the team(s) that had brought liquid crystals to the brink of commercialisation to be disbanded.

The corporate management, on the other hand, was distinctively 'unwilling' to comprehend such a concept. It was not that it lacked the ability to do so, it was just much more convenient for its own purposes if it 'pretended' it could not. Once RCA had decided it was not prepared to take the risk of putting the technology into production it had signalled that for all intents and purposes it had turned its back on the technology for good.

What Heilmeier does not reveal in his account of RCA's reluctance to develop LCDs was that the firm started to suffer from serious management problems from the late 1950s onwards, which eventually forced the firm to make a complete withdrawal from the consumer electronics industry. The process of decline started with RCA's near disastrous development of its own colour TV system. Graham (1986) documents how after a promising start, RCA's own system was eclipsed by a rival system developed by CBS which was subsequently approved by the US Federal Communications Commission (FCC).[29] The FCC's approval of the CBS colour standard in 1951 on the grounds that its technical performance was superior to RCA's was a bitter blow for the latter. The decision propelled RCA to redouble its efforts to radically improve its electronic colour television system and establish it as the new standard. RCA remained highly optimistic about the potential for its colour television system because despite the FCC's approval of the CBS system, it was incompatible with the then current black and white system. As the need for compatibility between the black and white and emerging colour systems became more urgent, the FCC was forced to re-examine the issue.[30]

After a two-year long consultation period involving some 200 engineers from ninety-one firms in the industry, the FCC proposed a compatible colour television system that was not dissimilar to the one originally proposed by RCA in 1949.[31] The FCC eventually approved the new system in December 1953.[32] The approval of the new system gave RCA its window of opportunity. With the FCC's work completed, RCA wasted no time in promoting its colour TV system in the market place. It was more aggressive in getting its system established than any of the other firms in the industry. Much of this had to do with the fact that it had invested an enormous amount of its own capital, some $65 million, which it obviously did not want to see disappear down a black hole.

Although commercial colour broadcasting commenced in January 1954, the market was not quite ready for RCA's colour television system. Both the consumer and the television manufacturers were quite happy with the black and white set. The consumer showed no sign of discontentment with the quality of the television picture and the manufacturers received a reasonable financial return. The stability of the black and white television set market forced RCA's competitors such as CBS, General Electric, and Zenith to withdraw their own colour models from the market.[33] Quite unexpectedly, RCA found itself having to bear the cost of maintaining the colour system on its own.[34]

Having invested so much capital and effort in developing its colour television system and getting it established in the market, RCA could not afford to simply walk away from the situation it was plunged into. All it could do was to sit tight and wait for the market to change. Whether it liked it or not, RCA had to wait seven years before the market decided it was time to move over to colour and it could start recouping its investment.

In many respects RCA's experience with its colour television system brought the firm down to earth with a bump. For the first time, RCA found that for all its capital resources and technological prowess, success was not automatically guaranteed. RCA's narrow escape with its colour television system badly shook the confidence that it had long enjoyed under its founder, David Sarnoff. Even with David Sarnoff still at the helm, top management knew only too well that the firm could ill afford to go through too many more such episodes. The biggest shock for management was that it was becoming less and less able to control and influence events. Its greatest concern was the increasing complexity and sophistication of the marketplace. Year by year it was getting more and more difficult to successfully commercialise new technology. This led top management to lose its nerve.

One of the immediate effects of top management losing its nerve was that it threw the firm's internal innovation process into considerable chaos. Graham (1986) notes that two major new technologies were 'killed off' once they had left the RCA Laboratories, transistor radios and *liquid crystals*.[35] Management's lack of ideas of what it wanted RCA to specialise in, and where it wanted to take the firm, led management to become totally indifferent to the potential of some of the technologies it had in its possession. This indifference emanated from the colour television saga. From 1954 onwards, there was a lack of continuity at the top. Management changes at the top did not give RCA the leadership it was desperately crying out for, or allow time for examination of RCA's fundamental problems.

The problem that was never addressed, and that plagued the firm for more than two decades, was the firm's continued desire to succeed at basic research in a range of technologies and product development at the same time. The colour television saga had exposed RCA's weakness in product development. In light of this, management proposed that the best way to insure RCA's future was to focus more on basic research. In 1954 RCA formulated its 'building block' strategy, which presumably was meant to gradually divert RCA away from an area it found extremely problematic, as its expertise in basic research grew and developed. Yet RCA continued to persist with product development. A vast quantity of managerial time was consumed in rectifying things when they went wrong, which they often did. Under such circumstances, it comes as no surprise that management became indifferent to certain technologies when it was so heavily tied up with other matters elsewhere.

When management did decide to commit itself to develop what it considered to be an 'exciting' new technology, the videodisc, it put all its eggs in one basket and got it all wrong. The constant technical problems and delays with the videodisc, which was finally launched on to the consumer market in March 1981,[36] pushed development costs beyond what RCA had budgeted for. Cost overruns put RCA's finances under considerable strain and left it short of capital to develop other major new technologies such as liquid crystals. Matters were not helped when RCA developed the Hayes & Abernathy syndrome. Scherer (1994) argues that RCA fell victim to this syndrome in 1976 when a new chief executive officer, with no technical background, took over the firm.[37] Worried by the firm's problems in getting a successful prototype of its videodisc into mass production, the new chief executive took the decision to delay mass production for five years.[38] Influenced by his 'detached' view of technology, the decision opened the way for Japanese firms to establish the videocassette recorder (VCR) in the market place way before RCA was in any position to compete with them head-on. The 'reasons' top management gave for RCA not developing liquid crystals were in the end no more than excuses to develop another technology which they genuinely expected had a far greater chance of restoring RCA's battered fortunes.[39]

Westinghouse

The years of poor management which led to RCA's eventual demise were replicated at Westinghouse during the 1960s and 1970s. In a personal account of the events Brody (1996) asserts the 'mismanagement' at Westinghouse, which led the firm to abandon a major technology and left an indelible mark of a negative nature on corporate America.

The first US patent for a TFT was issued in 1933 to JE Lilienfeld.[40] The second patent for a TFT was issued to Oscar Heil in Germany in 1935.[41] From what is known, no further work was carried out on TFTs until the 1950s with the emergence of the IC.[42] The urgent need for a more compact switching device led to the idea of MOSFET (metal-on-silicon field-effect transistor), and simultaneously, to the thin-film version of the same device, the TFT. The development of the semiconductor obliged several universities and many large firms such as IBM, General Electric, Westinghouse, RCA and Hughes to do research on MOSFET (especially MOS) and TFTs.[43] The outcome of all the research resulted in MOSFET being singled out as the technology of the future, the technology that would undergo future development and commercialisation. Once MOSFET had been selected the number of firms doing research on TFTs by the mid-1960s had fallen to just two, RCA and Westinghouse.[44] Towards the end of the decade RCA Laboratories too halted its research on TFTs on the grounds that it was a 'spent' technology, in other words, it was a technology that had failed to live up to its earlier promise of becoming a serious rival to MOSFET.[45] Westinghouse was thus left as the only firm in the TFT field.

Florida & Browdy (1991) claim that most firms abandoned this field because they had been unable to find a way of producing a cost-effective flat panel display.[46] This claim is very dubious as most flat panel technologies in the mid-1960s were still in an embryonic stage, so it is little wonder that firms could not produce a cost-effective flat-panel display! Brody offers a more plausible explanation which could account for why so many firms stopped their research on TFTs. MOS technology had emerged as the clear winner, so there was no point in firms' continuing in their quest to find another switching device after they had found one which met their requirements.[47]

As one of the country's leading producers of televisions and semiconductors in the early 1960s, Westinghouse was keen to develop new technologies in the areas where it excelled. One of the technologies identified as having great promise was TFTs. If they could be successfully developed they would be simpler, smaller and easier to manufacture than ordinary transistors.[48] Subsequently, Westinghouse set up a research team that included Brody to develop TFTs. Most of the internal financial support that helped sustain Brody's group came primarily from two sources, the consumer electronics division and the electron tube division. These divisions saw the development of flat panel technology as a way of restoring the firm's competitiveness in the television industry where it was losing market share.[49] According to William Coates, who once was an executive in the consumer electronics division, the plan was to make a wall hanging television,[50] which was precisely the same idea that RCA's LCD research group had.

The research on TFTs made important technical advances, but Brody's group was not able to go about its work with complete peace of mind. Florida & Browdy (1991) claim Westinghouse was typical of the big firms during the 1960s and 1970s, whereby R&D groups were required to generate funds from the operating divisions.[51] The weakness of this system was it turned scientists into 'salesmen' who had to give certain undertakings to obtain the funding for their projects. Brody's group was no exception and had to go through the necessary procedures of convincing the relevant divisions of the merits of their research. As noted above, most of their internal funding came from the consumer electronics division and the electron tube division. In return for their funding, the product divisions expected Brody's group, not unreasonably, to keep to tight deadlines, within budgets,[52] and develop a technology from which Westinghouse could benefit within a 'reasonable' length of time. The 'expectation' of the consumer electronics division was that Brody's team would develop a technology that was going to be cheaper than the cathode-ray tube (CRT).[53] Brody and his team seem to have found it easy to give these 'undertakings', be they verbal or written, it is of little consequence. When it came down to keeping to them in practice, the group found it very hard to deliver on their earlier promises which did not serve them well in the long term.[54]

Whilst Brody's group's failure to live up to its promises must have created tensions with management, Brody claims his group experienced trouble from an unexpected quarter, the integrated circuits division, which Westinghouse had set up in 1962. He argues that the integrated circuits division complained persistently to the laboratory's management concerning the work on TFTs. The former was of the opinion it was equivalent to pouring money down a black hole, and this put unnecessary pressure on Brody's group to constantly justify its existence and maintain its search for funding.[55] Brody saw this as unwarranted interference, but it looks as though it was a ploy to divert attention from problems that were brewing in the integrated circuit division's own backyard. In the mid-1960s, the competitive pressures of the semiconductor industry started to take their toll on Westinghouse as more innovative firms such as Fairchild, Texas Instruments, and Motorola began to make their presence felt in the market.[56] Westinghouse suffered a loss of market share, and with it a major loss of revenue. The firm's problems were compounded by its loss of market share in televisions. Losses in these two sectors placed an almost intolerable burden on the firm's financial resources. As part of a wider programme to restore profitability, the integrated circuit division was closed down in 1968.[57] For the corporate management and the product divisions, what really mattered at the end of the day was the bottom line, profitability. Brody is at pains to point out the divisions that gave his group substantial financial support were no different from the other divisions of Westinghouse, they were 'short-term oriented divisions with quarterly profit responsibilities'.[58]

With the integrated circuit division gone, and with it one of his most persistent critics, Brody found keeping the support necessary to keep the group's work on track suddenly a lot easier. The turnaround in the group's fortunes was sufficient to allow it to grow, and it subsequently became known as the 'Thin Film Devices' group, of which he became manager. Just as other semiconductor firms who had been working on TFTs were in the process of abandoning the field, as it was widely believed TFTs were of no use, the Thin Film Devices group found some useful applications for them. One such application was an aircraft power-control circuit, which for the first time combined logic and power devices in a single integrated network (long before the semiconductor produced such circuits).[59] Another application was the development of a 400-V high-input-impedance multiplexer circuit with a 120-dB common-mode rejection, which could replace high-cost mercury wetted relays.[60] This particular application was part of an exercise to find applications for which conventional semiconductor components and circuits were not completely suitable, whether it was in terms of performance, cost or availability.[61] At this stage the Hayes & Abernathy syndrome did not seriously affect Westinghouse's activities. The Thin Film Devices group's work was deemed sufficiently important by the product divisions to warrant their continued support. It was, however, just a matter of time before the syndrome finally hit the firm.

A major point, which needs to be made here, is that the work on TFTs prior to 1968 was carried out before any serious consideration had been given for their use in displays.[62] In conjunction with work on TFTs, Westinghouse had also carried out research on one of the other major flat panel display technologies, electroluminescence. The importance of this research, however, was not readily apparent until Brody hit upon the idea of using TFTs in a display in 1968. His team was successful in their very first attempt to make a TFT device switch on and off an electroluminescent display. This was the birth of what is now known as the 'active-matrix' addressing system, used in the third generation of LCDs produced by all the major Japanese and South Korean firms in the industry. The term 'active-matrix' (AM) did not appear until the publication of a paper by Brody in 1975; like the technology it is now in universal use.

What is an active-matrix addressing system? The image produced on a conventional flat panel display reaches it via rows and columns of electrodes. The intersections of these electrodes form a grid of picture elements (or pixels). A pixel is turned on and off when a voltage is present in both the row and column electrode. This technique, known as multiplexing, is adequate for small sized LCDs, but even then it cannot produce an image that is superior to the one produced by a CRT. When this design is used in a moderately sized LCD, where there are more pixels, as the voltage passes through the electrodes, it also affects the individual liquid crystal molecules. The molecules that should be turned off are slightly turned on, and instead of getting a clear black and white image, an image of various shades of grey is produced. Furthermore, an LCD's ability to produce high quality fast moving pictures is restricted by multiplexing. The response time of an LCD is proportional to the voltage applied. Since multiplexing limits the voltage that can be applied the LCD cannot produce high quality fast moving pictures. These can only be produced by an LCD when the response times of the liquid crystal molecules are exceptionally rapid. The active-matrix design avoids these problems by placing a separate semiconductor switch, a TFT, at each pixel. The virtue of the TFT is that it can transfer and store enough voltage to quickly switch a liquid crystal pixel from light to dark, to create a very sharp image with no blurring. A pixel with its own TFT, for example, can be switched on and off without affecting any of the surrounding ones. One further advantage of this design is that the TFTs are deposited on a glass substrate, enabling a display to be lit from behind thus improving the image it produces.[63] In what is regarded as a classic paper within the electronics industry, Brody reported in 1973 that he had successfully used his active-matrix addressing design in an LCD.[64]

Even though the Thin Film Devices group could demonstrate that it continued to make important advances, Brody claims that in the intervening years, between the time he hit upon the idea of using TFTs in displays and the publication of his 1973 paper, life for the Thin Film Devices group was 'uncomfortable'. The main thing that made him so annoyed was the degree of scepticism, which came from a number of areas, for example, those who believed the CRT would reign supreme indefinitely, and the laboratory's management. According to Brody, one authority on CRT technology questioned the value of the group's work by asking, 'Who needs a flat panel? The CRT is so good, nothing will ever replace it, and besides, you don't stand a chance of building a manufacturable flat display, lots of people have tried and failed, why try again?'.[65] The laboratory management were equally unhelpful when they said, 'We let you get away with TFTs so far because you managed to get divisional support, but we are certainly not going to fund such a crazy project',[66] referring to the development of a flat panel display.

Brody became even more convinced within himself about the value of TFTs after he had immersed himself in the literature on flat panel technology. To his great surprise he found that innumerable attempts had been made to produce a flat panel display without success over a twenty-year period. It was not just the sheer number of attempts that took his breath away, it was also the number and variety of the electro-optical materials which had been proposed and worked on.[67] The common link that Brody identified amongst a very high proportion of these early attempts, which had ended in failure, was the lack of attention paid to finding an efficient mechanism for distributing a picture/image to the pixels.[68]

One of the more useful suggestions given by the laboratory's management to the Thin Film Devices group was that if they wanted to continue their research on TFTs, they should explore the possibilities of government funding. It was a suggestion that in time gained the group a very useful source of funding. In terms of using the active-matrix addressing system for a flat panel display, the most likely technology in 1968 was electroluminscence as liquid crystal technology was still in its infancy. By 1971 the situation had changed significantly and liquid crystals had emerged as a real contender to electroluminescence. In the same year, Brody's group received $60,000 from the Air Force and some additional funding from the Army. The funds from the Air Force were to build a 6×6 inch AM-LCD, and for the Army an active-matrix electroluminescence display of the same specifications. The challenge the group then faced, now it had the contracts it needed, was to create its own production technology, that is, they had to design and make their own active-matrix fabrication equipment.[69] Because the technology they were working with was at the cutting-edge, there were no indigenous suppliers in existence. All the group could manage to focus on at this stage was to find a method of producing small quantities of active-matrices suitable for use in prototypes, and master the art of liquid crystal alignment, filling, and sealing.[70] The progress of Brody's group was quite remarkable. In 1973, approximately two years from the date that it had secured the funding from the DOD, Brody's group had built the world's first operative AM-LCD.

Despite having built a functioning AM-LCD, the Thin Film Devices group found itself once again experiencing difficulties. The source of the problem this time was the Department of Defense (DOD). A committee called the Advisory Group on Electron Devices (AGED) had the responsibility for reviewing the group's funding. For Brody, the outcome of the review was going to be unfavourable before it had even taken place. The committee was made up of 'experts' from the semiconductor industry who had already decided that TFT technology had no future. When they learned that the Army and Air Force were actually funding it, the committee immediately halted the release of any future funds on the pretext that it was a 'waste of government funding'. While the funding from the Air Force was terminated, the funding for the Army was allowed to continue, albeit under very close scrutiny from AGED.[71] Brody puts the blame for this 'fiasco' at the door of George Heilmeier. He then held a highly influential position within the DOD and was opposed to research on TFTs. Brody alleges Heilmeier's 'prejudice' against TFTs originated from his time at RCA where it was apparently found that TFTs did not work.[72]

Perhaps not surprisingly, the actions of the DOD had an immediate impact on Westinghouse's corporate management. It came to the obvious conclusion that if the DOD had not considered the research worthy of long-term support, AM-LCD research was indeed equivalent to pouring its money down a black hole. But it was some considerable time before it was finally able to bring a complete halt to the firm's activities in the area. Whilst corporate management had a negative view of the AM-LCD, it had a noticeably more positive attitude towards the AM-electroluminescence display which enjoyed significant DOD support.

Because of this support the corporate management 'advised' the Thin Film Devices group to focus their efforts on this technology. With the DOD's active support, the electron tube division proceeded to supplement the DOD's funding with quite substantial divisional funds.[73] The willingness of specific divisions to support the research of the Thin Film Devices group, rather than the corporate management, shows Westinghouse suffered from a strange paradox. The divisions displayed a far stronger awareness of the need to develop new technologies than the corporate management ever did, even though both parties were concerned about short-term profitability. For the corporate management, profitability remained their top priority, but for the divisions it did not take priority over absolutely everything else. So long as their profitability was not under serious pressure, funding basic research and product development was seen as an important part of a division's activities. Attitudes quickly changed though at the sign of any major trouble. More often than not, as soon as divisions found themselves in any serious trouble, any funding which they provided quickly ceased.

The combined funding of the DOD and electron tube division kept the Thin Film Devices group in business at least, even at the price of having to sacrifice work on what they believed was a more promising technology. So long as the funding continued, work on active-matrix technology could continue, but in 1976 the Thin Film Devices group received a fatal blow. The fatal blow came from the closure of the electron tube division and the consequent loss of a substantial proportion of funding. The immediate cause was Westinghouse's continued loss of market share in the television industry, which by the mid-1970s had dwindled to 3 per cent in the black and white market and to just 1 per cent in the colour market.[74] This loss of market share was symptomatic of a much deeper malaise at Westinghouse; it had lost its competitiveness in areas where it had once been strong and this precipitated a change of management at the top. In its wisdom, the new corporate management decided to axe some of the unprofitable areas of the firm's activities, including televisions, which they decided to stop making altogether, and a semiconductor division that had once supported Brody's work.[75]

Very gradually, Westinghouse was beginning to look less and less like an electronics firm. Its 'technology strategy' which it appeared to have a few years earlier had completely disintegrated. As mentioned earlier, when the consumer electronics division started to give financial support to the Thin Film Devices group, it had in mind a very good idea of what the technology might one day be used for, wall hanging televisions. By the mid-1970s, Westinghouse was still doing research on TFTs, but it had no concept of what it could do with the technology once its television operations had been closed down. Whereas it had once had a solid grasp of the 'innovation market', like RCA Laboratories did, the poor financial health of Westinghouse caused it to gradually vanish from the corporate mindset. The need to sort out its financial problems did not allow the firm the opportunity to step back and take a look at the potential possibilities of what it might be able to do with TFTs, and assess what other applications they might otherwise be used in besides wall hanging televisions. Any search it may have conducted for another 'competitive' technology, which had the potential to restore the firm's dwindling fortunes, would have been hindered by two key factors. Firstly, the firm's absence from the computer indus-try and terminal business,[76] and secondly, the fact that the earlier idea of constructing a wall hanging television had fallen from grace. What had brought the latter into serious disrepute was the failure of Westinghouse's earlier attempt to build a flat panel display, an electro-luminescent-ferroelectric (EFL) display, in the early 1960s.[77]

Once the axe had fallen on the electron tube division the demise of the Thin Film Devices group seemed inevitable. Much to its surprise it was thrown a temporary lifeline by the DOD. The group received two contracts, one from the Army for research into manufacturing processes, and the other from the Navy to restart work on AM-LCDs. In spite of all its efforts and achievements, the future of the group was never really in doubt, it simply did not have one. The new manage-ment was determined to have its way and was not going to allow any-thing to blow it off course. According to Brody, a manager with an accounting background replaced his former Business Unit Manager, who had been a competent scientist. Once the new manager had his feet firmly under his desk he set about bringing the group's work to a complete end, but it was not a straightforward task and the whole process dragged on for two years.[78]

The problem was the Army; the contract was a large one and a number of review committee meetings had to be held before it could be finally concluded. After Westinghouse had completed its contrac-tual obligations, the firm took the opportunity to cancel all future work. The basis of Westinghouse's decision to halt the work that it was doing for the Army and to disband the Thin Film Devices group in 1978 was that it was 'unprofitable'.

Brody's project was allegedly a 'high risk' one, according to the cor-porate guidelines set down.[79] On the basis of the criteria for this cate-gory of project, profitability was then analysed with the aid of a computer. The results of its analysis suggested that if work had been allowed to continued the best return on investment (ROI) Westing-house could have expected was 13.2 per cent, when the hurdle rate was set at 13.4 per cent.[80] It is hard to say with absolute certainty whether or not this was a 'manufactured' figure, one that was used to bring the work on TFTs to a halt. On the one hand, Westinghouse had already invested several million dollars by this date in TFTs, and it would have taken a very sudden and complete change in the firm's thinking to reach the conclusion that it could afford to see all its investment go to waste on such flimsy grounds. On the other, it is quite conceivable that Brody's new Business Unit Manager, saw that his group's work was not going to be profitable in the short term and decided (with the blessing of corporate management) that the work on TFTs had to be stopped to save Westinghouse money, irrespective of how much the firm had invested in the technology in the past. In light of how important short-term profitability was to Westinghouse, the odds are that it was the actions of Brody's new Business Unit Manager which finally brought the curtain down on the firm's research on TFTs. Westinghouse had finally been struck by the Hayes & Abernathy syndrome.

Once the Thin Film Devices group had been disbanded and Westinghouse had made it absolutely clear that it was no longer interested in TFT technology, Brody found himself out on his own. Determined not to give up on something which he passionately believed in and had spent a number of years developing, Brody decided to exploit TFT technology on his own. Little did he realise when he made his decision what an uphill struggle it would be. The biggest single factor that was to make his task so difficult was the actions of Westinghouse. Firms and venture capitalists alike, all suffered from the 'sheep mentality', because they all duly followed each other without hesitation. The reluctance of venture capitalists and firms to back TFT technology was so prevalent that Brody was turned down on more than eighty occasions by such firms as IBM, Hewlett-Packard, Texas Instruments, and Matsushita.[81] They reasoned if active-matrix technology was so promising, why did Westinghouse abandon it? The perception the venture capital community had of Westinghouse was that it had spent years developing a 'promising' technology and had got nothing back in return for it.

The biggest factor to weigh on the minds of the venture capitalists and US firms was that by now Japan had caught up in TFT technology. Japan's performance in the semiconductor industry was more than enough for them to not even begin to contemplate competition. Once corporate America had lost interest in TFT technology, it would be more than a decade before it took a renewed interest, and by then it was too late.

The role of government

In comparison to its role in semiconductors, the government's role in the development of LCDs was absolutely minuscule. The only funding which the government channelled into LCDs was in the form of two military contracts. The first contract that Brody won came from the

Air Force in 1968 and the second from the Navy in the mid-1970s. When the size of the first contract (worth $60,000) is compared with what the government spent on military research annually, it is a drop in the ocean. In 1962 an article in *Electronics International* quoted a government publication,[82] where a senator is reported to have said that, 'A total of $9 billion is currently expended by the Federal government for research, development, testing and evaluation. Not less than one-fourth of that sum is spent in the electronics field alone'.[83] There is no evidence to suggest otherwise that the government tried to 'target' LCDs during the 1960s and 1970s. As far as displays were concerned generally, by the early 1960s there was mounting evidence which indicated that in the coming years there would be significant demand for large displays from both the military and civilian sectors.[84] One estimate made in 1963 claimed the display market was already worth $200 million.[85] There was, however, no indication which technology would eventually emerge as the dominant one.

The rationale behind the Air Force's decision to award Brody a contract in 1968 to develop AM-LCDs has to be seen in a wider context. At the top of the government's priorities was the need to provide the military with the best available technology at its disposal, a major necessity considering the country was caught up in the midst of the Cold War. The Air Force had a need for light and reliable displays that could easily translate data into information that would not be misinterpreted.[86] It was therefore interested in funding research into any display technology that might lead to the development of its 'ideal' display. There were a number of display technologies around at the time and LCDs were just one of the options the Air Force was keen to explore. The contract from the Navy in the mid-1970s was an indication of how far the LCD had come in comparison to other more mature display technologies, and growing recognition of its emerging flexibility.

Summary

In this chapter we have pinpointed where the path for understanding the development of the LCD industry begins, and identified the fundamental cause which led to two highly innovative US firms allowing LCD technology to slip through their fingers, the Hayes & Abernathy syndrome. In addition, what is also very evident is a major weakness in the country's NSI: it discouraged firms from learning about a major capital intensive technology, and the venture capital community's apparent lack of willingness to invest substantial amounts of long-term capital to put the technology into production. In sum, it was a combination of the Hayes & Abernathy syndrome, and this major weakness in the NSI which were the principal factors that led the US to give away its lead in LCDs to Japan, a lead that it has never regained.

Notes

1. Collings, PJ (1990) *Liquid Crystals: Nature's Delicate Phase of Matter*, Bristol: Adam Hilger, p. 25.
2. *Ibid.*
3. *Ibid*:25–26.
4. *Ibid*:26.
5. *Ibid*:25–27.
6. *Ibid*:27.
7. *Ibid.*
8. *Ibid.*
9. *Ibid*:28. & Kelker, H (1973) 'History of Liquid Crystals', *Molecular Crystals & Liquid Crystals*, Vol. 21, No. 1, p. 7.
10. Kelker, 1973:7.
11. Quoted in *Ibid.*
12. Collings, 1990:28.
13. Kelker, 1973:9.
14. Heilmeier, GH (1976) 'Liquid Crystal Displays: An Experiment in Interdisciplinary Research that Worked', *IEEE Transactions on Electron Devices*, Vol. ED-23, No. 7, July, p. 781.
15. Kelker, 1973:17–31.
16. Heilmeier, 1976:781.
17. Collings, 1990:31.
18. *Ibid*:32.
19. *Ibid.*
20. *Ibid.*
21. 'Screen Technologies: Too soon to write off the cathode ray tube', Mobile Computing Survey, *Financial Times*, 26 January 1994.
22. Heilmeier, 1976:781.
23. The main figures in the group were George Heilmeier, Louis Zanoni, Joel Goldmacher, Joseph Castellano, and Lucian Barton. Castellano, JA (1988) 'Liquid Crystal Display Applications: The First Hundred Years', *Molecular Crystals & Liquid Crystals*, Vol. 165, p. 390.
24. Heilmeier, 1976:784.
25. *Ibid.*
26. *Ibid*:785.
27. *Ibid*:784–5.
28. 'RCA liquid crystals in watch display', Electronics newsletter, *Electronics International*, 5 July 1971, p. 17.
29. Graham, MBW (1986) *RCA and the VideoDisc: The Business of Research*, Cambridge: Cambridge University Press.
30. *Ibid*:62–65.
31. *Ibid*:64.

32. *Ibid*:65.
33. *Ibid*.
34. *Ibid*.
35. *Ibid*:86.
36. *Ibid*:213. The eventual loss incurred by RCA as a result of the videodisc's failure was $580 million.
37. Scherer, FM (1994) 'Competing for Comparative Advantage Through Technological Innovation' in O Granstrand (ed.) *Economics of Technology*, London & Amsterdam: Elsevier Science, North-Holland, p. 348.
38. *Ibid*.
39. RCA's finances and reputation were severely damaged when it was forced to withdraw from the computer industry in the early 1970s (Graham, 1986:144).
40. Brody, TP (1984) 'The Thin Film Transistor: A Late Flowering Bloom', *IEEE Transactions on Electron Devices*, Vol. ED-31, No. 11, p. 1614.
41. *Ibid*.
42. Brody, TP (1996) 'The birth and early childhood of active-matrix: A personal memoir', *Journal of the Society for Information Display*, Vol. 4, No. 3, p. 113.
43. *Ibid*:113–114.
44. *Ibid*:114.
45. *Ibid*:116.
46. Florida, R & Browdy, D (1991) 'The Invention that Got Away', *Technology Review*, 94, No. 6 (August–September) p. 47.
47. Brody, 1996:114.
48. Florida & Browdy, 1991:47.
49. *Ibid*:48.
50. *Ibid*.
51. *Ibid*:47–48.
52. *Ibid*:48.
53. *Ibid*.
54. *Ibid*.
55. Brody, 1996:114.
56. Florida & Browdy, 1991:47.
57. Brody, 1996:115.
58. *Ibid*:115.
59. *Ibid*:116.
60. *Ibid*.
61. *Ibid*.
62. *Ibid*:114–115.
63. This section has been taken from Florida & Browdy, 1991:45. The technical aspects of how the active-matrix design functions have been taken directly from the text, while other aspects have been substantially revised.
64. Brody, 1996:117.
65. Quoted in *Ibid*.
66. Quoted in *Ibid*.
67. *Ibid*:116.
68. *Ibid*.
69. *Ibid*:118.

70. *Ibid*:118–119.
71. *Ibid*:119–120.
72. *Ibid*:119.
73. *Ibid*:120. No figures are given.
74. Florida & Browdy, 1991:48.
75. *Ibid*.
76. Browdy, 1984:1618.
77. *Ibid*.
78. Brody, 1996:120–21.
79. *Ibid*:121.
80. *Ibid*.
81. *Ibid*.
82. The publication was *Coordination of Information on Current Research and Development in the Field of Electronics* (20 September 1961).
83. 'Our Growing Markets: An analysis of the present and a look into the future', *Electronics International*, 5 January 1962, p. 49.
84. 'Large Displays: Military Market Now, Civilian Next', *Electronics International*, 25 January 1963, pp. 24–26.
85. 'Displays: $200 million', (under the heading 'Santa Monica'), *Electronics International*, 29 March 1963, p. 15. The market comprised predominantly of hardware and development contracts.
86. *Ibid* (under the heading 'Military Needs').

7
Supporting a New System of Innovation: Japan's Strategy in Liquid Crystal Displays

In this chapter, Japan's position in high technology industries and the strategy it used to build up a dominant position in the LCD industry are analysed. In the first section, it is shown that over the long term Japan's position in high-technology industries had steadily improved, largely at the expense of Western Europe and the US. From the perspective of Western Europe and the US, the various indicators show that Japan manoeuvred itself into a very enviable position. As discussed in Chapter 4, during the 1980s Japan's strength in high-technology industries manifested itself in its meteoric rise in the semiconductor industry. In the 1990s the country's continued strength in high-technology industries had manifested itself in its dominance of the TFT-LCD industry.

The second section starts with an examination of the digital watch industry in the 1970s. As stated previously in Chapter 1, it was turmoil in the US digital watch industry that dealt a devastating blow to the US LCD industry, but strangely left Japan's LCD industry virtually unscathed. In the previous chapter, it was concluded that the US gave away its lead in LCDs primarily because of a combination of the Hayes & Abernathy syndrome and a major weakness in its national system of innovation. The turmoil in the US digital watch industry ties in very well with this as it provides additional evidence of why the venture capital community proved so reluctant to invest in LCD technology. It can only be a matter of conjecture, but it seems that the venture capital community must have reasoned that it was simply not financially viable to invest in an industry that had fallen into serious decline. The chances of US LCD firms, which had been severely weakened financially by the turmoil in the US digital watch industry, competing with Japanese LCD firms were negligible at best. After all, if US firms were finding it difficult to compete head-on with Japanese firms in semi-conductors, an industry where the US had long been dominant, US LCD firms had no chance of competing against the very same Japanese electronics firms in an industry where Japan was now dominant.

Besides the very poor financial state of US LCD firms, there was also another important factor that made the venture capital community very reluctant to invest in LCD technology, which Brody (1984, 1996), Florida & Browdy (1991), and Tyson (1992) fail to mention.[1] The strategies of US and Japanese LCD firms were very different from each other. As we shall see later, from the early 1970s Japanese firms made a determined effort to develop LCD technology by setting up research programmes, and establishing long-term relationships with other firms interested in jointly developing LCD technology. US firms, on the hand, did nothing of the sort. Even if there had been no turmoil in the US digital watch industry, one wonders how long it would have been before US LCD firms would have succumbed to competition from Japanese LCD firms. How could US firms expect to compete in a technology in which they had few intentions of investing, while Japanese firms were?

The new technological challenge

Japan's assault on the semiconductor industry during the 1970s and early 1980s is a remarkable example of industrial catch-up, but it cannot be viewed in isolation. In fact it was part of a much bigger phenomenon that saw Japan emerge as the new 'workshop of the world' after the mid-1970s. The shock waves of this seismic change were not felt for some time, although the early warning signs in Western Europe and the US indicated that Japan was going to become a major competitive threat to them in the future. By the early 1980s, Japanese firms had captured significant shares in overseas markets in a number of key industrial sectors ranging from cars to machine tools and electronics, with seemingly apparent ease. This changed the 'old' post-war *status quo* forever. Major industrial sectors in Western Europe and the US started to face competition of a severity which they had never previously experienced. An additional problem for the US was that it found its technological hegemony was being seriously challenged for the first time. Japan's rapid rise to prominence in semiconductors indicated that its success had been built on very solid foundations, and that no sector, not even high technology ones, could ever be considered 'immune' from Japanese competition.

The depth of Japan's strength in high-technology industries, and the simultaneous decline experienced by Western Europe and the US are substantiated by a number of important indicators. As a proportion of total manufacturing output of the advanced industrial nations, the percentage of technology-intensive products has risen steadily over the last two decades. In 1990 high technology production accounted for approximately 20 per cent of Western European manufacturing output, 30 per cent for the US, and 35 per cent for Japan.[2] While these headline figures are relatively comparable with each other, other figures tell a very different story. Between 1980 and 1990 the shares of Western Europe and the US in global high technology production, as defined by the OECD which uses a 'narrow' classification, as opposed to the Guerrieri & Milana classification which is much broader, declined by approximately 17 per cent and 11 per cent, respectively, while Japan's share increased by 59 per cent.[3]

This fall in production was not the only setback to hit Western Europe and the US, they have also seen their global shares of high technology exports (using the Guerrieri & Milano classification) decline. Between the periods 1970–73 and 1988–89, Germany's share fell from 16.79 per cent to 12.52, a decline of 4.27 per cent; France's share fell from 7.22 per cent to 6.80 per cent, a decline of 0.42 per cent; the United Kingdom's share fell from 10.12 per cent to 7.64 per cent, a decline of 2.48 per cent; and the US's share fell from 29.54 per cent to 20.64 per cent, a decline of 8.90 per cent.[4] Japan's share of high technology exports, on the other hand, showed a corresponding rise. Its share increased from 7.07 per cent to 16.01 per cent, an increase of 8.94 per cent.[5] One interesting point Tyson makes is that Japan's share continued to rise, and the US share continued to fall during the mid-1980s, after there had been a major appreciation of the yen against the dollar.[6] When examined more closely, the major factor which has led to the change in the terms of trade in favour of Japan has been the long-term competitiveness of its electronics industry. In Figure 7.1, the trend is clear to see, from the early 1970s through to the late 1980s Japan's global share rose steadily, while the shares of Western Europe and the US reveal a distinct pattern of decline.

The disparities in the technological performances of Western Europe and the US vis-à-vis Japan mirrored by the sustained loss of market share, created alarm about the opening up of a technology 'gap', especially in Western Europe. The origins of this alarm date back to the 1960s when Western Europe feared a technology gap might develop

Country	1973–79	1979–82	1985–88	1988–89
Japan	9.6	17	23.4	24
US	28.9	23.8	19.2	18.3
Western Europe	44.7	37.4	30.8	28.6

Figure 7.1   Shares of world trade in R&D-intensive electronics
Source: Adapted from Tyson, 1992:25[7]

with the US. The major catalyst for the concern was the publication of *The American Challenge* by J. Servan-Schreiber in 1967, which centred on worries about Western Europe's growing dependence on American technology, then epitomised by the overwhelming dominance of IBM in the computer industry.[8] Patel & Pavitt (1987) point out these fears were subsequently proved to have been misplaced. Over the following decade, the industrial performance of Western Europe was better than that of the US, whether in terms of output, production, exports, or of technology as reflected in R&D expenditures.[9]

The digital watch industry

On the surface, the changes which the digital watch industry experienced during the 1970s appear to be of no significance to anyone today as watches are low-value products and are not that technologically sophisticated. However, the changes which took place in the industry radically altered the structure of both the US and Japanese LCD industries, which created an opportunity for the Japanese industry to exploit LCDs in the absence of any significant overseas competition. The digital watch was important to both US and Japanese LCD industries, as it was one of only two sources of demand for the twisted-nematic (TN) LCD. Watches accounted for 57 per cent of the total volume and calculators accounted for most of the rest in 1977.[10] The increasing popularity of the LCD digital watch provided an excellent foundation for the development of LCDs, for example, in 1973 the US LCD market was worth $1.6 million, but had increased to $40.9 million by 1980.[11] It was the Japanese industry though that eventually benefited from this growth rather than the US industry.

One major point about the US and Japanese digital watch markets which cannot be emphasised enough is that they were very distinct from one another from the beginning. As is usually the case with any fast growing industry in the US, US firms, in Schumpeterian terms, 'swarm' into it in pursuit of 'supernormal profits', and the digital watch market was no different. Firms swarmed into it at a frenetic pace as the opportunities for them dramatically increased. In Japan though, the digital watch market did not experience anything like the scale of swarming the US did. In actual fact there was virtually no swarming at all. The only plausible explanation for this is the *keiretsu* structure of the Japanese economy. With the creation of approximately 20 *keiretsus*, the scope for swarming has been severely limited.

When Japanese firms do swarm into a fast growing industry they do so with the express intention of being there for the long-term and not just for 'easy' short-term profits. In certain sectors, for example, televisions, Japanese firms swarmed in 'considerable' numbers. What appears to have deterred the majority of firms from swarming into the digital watch industry was the lack of scope for growth. When we talk about a considerable number of firms in Japanese terms, this normally means a figure of approximately ten. The term 'considerable' takes on a different dimension when it is applied to the US. When firms swarm into a particular sector, the incumbent firms often find they are joined by numerous other firms from different sectors who are attracted by the lure of supernormal profits. The number of domestic firms in a rapidly growing sector can easily surpass twenty or more. The end result of this massive swarming is that once the supernormal profits have dissipated, the industry becomes so unprofitable that the great majority of firms are forced to exit, paving the way for foreign firms to fill the vacuum.

Throughout the 1970s, Japan tightened its grip on the development and production of both the digital watch and calculator. At the start of the decade the US and Japan were, to all intents and purposes, 'equal' in watches, but in calculators Japan had a considerable edge over the US.[12] By 1970 Japan had managed to capture between 79–85 per cent of all electronic calculator sales in the US.[13] US firms responded quickly and positively to the Japanese challenge. Firms such as Texas Instruments, Hewlett Packard and National Semiconductor, along with a host of semiconductor firms entered the calculator market in the expectation of making supernormal profits. Texas Instruments entered the industry in 1972 and launched its first three models of calculators in the September of the same year, and Hewlett Packard adopted what was then a popular strategy of backward integration, which took the firm into the highly competitive semiconductor industry.[14] The overall effect of this surge of US firms into the calculator industry was that by 1974 Japan's share of the US electronic calculator market had declined to only 25 per cent.[15]

The aggressiveness with which Texas Instruments responded to the challenge posed by Japanese firms, helped turn it, within a year of its initial entry, into the industry's largest firm,[16] and later its most profitable.[17] The sudden flush of success which US firms experienced against the competition from Japan did not last for very long. As the average price of a calculator declined from $100 in the latter part of 1972 to $35 in 1975, the US 'challenge' began rapidly to disintegrate.[18] The plunge in prices led many of the firms which had entered the calculator industry just a few years earlier to sustain massive financial losses, forcing a huge swathe of them to leave the industry altogether. Partly because of their *keiretsu* affiliations and much greater efficiency in production, Japanese firms rapidly recaptured the ground which they had lost to US firms prior to 1974. The exodus of US firms from the mass market created a vacuum which was promptly filled by Japanese firms. Once the US firms had left they never returned en masse, and Japanese firms have continued to dominate the industry ever since.[19]

For Japan, calculators proved not only a valuable source of demand for LCDs, but they prevented the country from having to put all its eggs in one basket, and rely on a single product, the digital watch. Although calculators underwent enormous development nothing of critical importance occurred in this sector of the electronics industry that has had a long-term impact on LCDs. As it turned out, it was the digital watch that was to provide all the fireworks.

One of the first indications that the digital watch was on the verge of becoming a commercial reality was in 1970 when it was reported that a low-powered MOS (metal-oxide semiconductor) chip could be used in a watch, but it was stressed it would be some time before it would reach the mass market.[20] One of the main barriers which prevented it from being mass-produced was the very high production costs, for example, Seiko sold one model of watch for $650. Costs were projected to fall over time as firms became familiar with the new technology and competition increased, one forecast predicted the simplest models would eventually cost between $10 and $20.[21] The other barrier which slowed the progress of the digital watch to the mass market was the variable quality of the display devices.

The two display devices used in digital watches were the LCD and the light-emitting diode (LED). At the beginning of the 1970s, both these display devices suffered from major drawbacks, but it was the LED which had the early advantage over the LCD. The strengths of the LED were that it was very reliable, it could function over a wide temperature range, it had a very long life span, and it generated its own light (usually red). Its weaknesses were that it was power-hungry, the cost of the LED soared as the size of display increased due to the expensive light-emitting semiconductor technology, and the quality of the display was very poor in bright sunlight.[22] The strengths of the LCD were its very low-power consumption, it was cheap to produce almost regardless of size, and the quality of the display in bright sunlight was markedly better than the LED. Its weaknesses were its limited life span (approximately 10,000 hours) and its inability to function over a wide temperature range.[23] Although any user of an LED watch had to push a button to confirm the time as it consumed too much power to show a continuous display, its reliability maintained demand at a high level until the mid to late 1970s. In hindsight, this was to prove to be a major watershed as it was during this period that the performance of the LCD watch saw a dramatic improvement, triggering the rapid demise of the LED watch.

In the US market the demand for LED watches was much stronger than for LCD watches.[24] The buoyancy of the digital watch industry was reflected by the number of firms present. In 1975 one report claimed the number of firms had reached 40, which does not seem too wide of the mark.[25] This headline figure appears to have included those selling digital watches as well as those producing and supplying the various components, modules and displays. Figure 7.2 gives the names of 32 firms which are known to have entered the digital watch industry.

The 'traditional' watch firms, such as Timex, soon found their traditional territory being swamped, especially by large numbers of semiconductor firms because of the huge opportunities. To the semiconductor firms, the advent of the digital watch was a major new source of demand for integrated circuits, and almost every semiconductor firm wanted to gain a slice of this growing market. Like the calculator industry, the digital watch industry appeared to present firms with yet another opportunity to make supernormal profits. Although if any firms made supernormal profits, they only did so for a period of perhaps four years. The digital watch was first introduced in 1972,[27] and like every new electronic product that

American Microsystems Inc	Hamilton Watch Co	Kylex	Optel Corp
Beckman Instruments	Hamilton Technology Co	Microdisplay Systems	Princeton Materials Science
Benrus Watch Co	Heuer Time & Electronics Corp	Microelectronic Systems Corp	Pulsar division of HMW Industries
Bomar	Hughes Microelectronics	Microma Universal	RCA Solid State Divison
Computer Time Corp	Integrated Display Systems	Micro Power Systems Inc	Solid State Time
Croton	Intel	Motorola	Texas Instruments
Fairchild Camera & Instrument	International Liquid Crystal Corp (ILIXCO)	National Semiconductor	Timex
General Time Corp	Intersil	Novus	Walthum

Figure 7.2   US firms in the digital watch industry
Source: Numagami, 1996:143–146; *Electronics International* (various copies)[26]

comes on the market it was initially very expensive. Within four years of its appearance on the market prices had collapsed, following a pattern highly reminiscent of the calculator industry.

In stark contrast to the US market, the Japanese market was initially dominated by the quartz-crystal watch. In 1977 Seiko claimed the quartz-crystal watch outsold the digital watch (in this case the LCD watch) by a ratio of between 4:1–5:1.[28] Demand for the LED watch remained very low and this was met largely through imports from US firms.[29] Ricoh was the only major firm that opted to use LEDs for use in its digital watches.[30] Figure 7.3 illustrates exactly how the levels of demand for the LED watch and LCD watch varied between the US and Japanese markets.

As Figure 7.3 indicates, the US market between 1977 and 1978 saw a distinct shift in the pattern of demand away from the LED watch towards the LCD watch. Even though the demand for the LED watch remained quite robust, it failed to prevent a radical shakeout of the industry. The seeds of the shakeout had been sown a couple of years earlier, when the demand for the LCD watch began rapidly to increase. The rise in demand for the LCD watch in 1976 had

Consumption of digital watches in the US and Japan

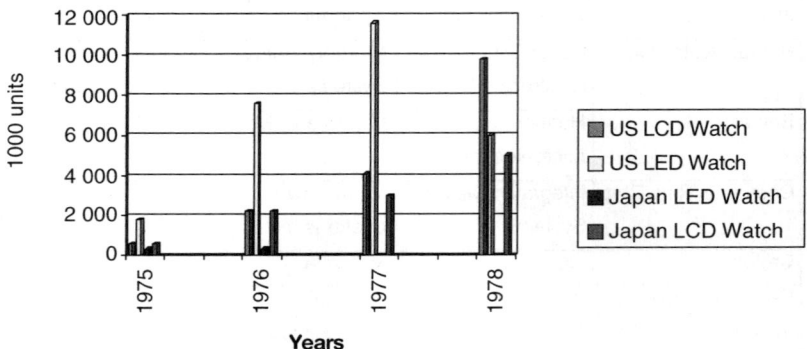

Figure 7.3   Consumption of digital watches in the US and Japan
Source: Table compiled from data in Numagami, 1996:138

prompted fears that in the subsequent year there might be a short-age of LCDs,[31] but these were later shown to have been misplaced.[32] Although there were plenty of LCDs being produced, the price of LED watches in the US market fell precipitously due to intense competition in 1977, putting a great many of the firms under enormous financial pressure. Virtually every firm in the industry was affected by the collapse in prices, because as the price of LED watches plummeted, it dragged down the price of LCD watches as well. This was a sure sign that the process of 'creative destruction' had begun to take root.

The intense financial pressures and the decline in the demand for the LED watch forced many of the firms to leave the industry altogether. Many firms were simply left without the necessary resources or time to make a successful transition from the LED watch to the LCD watch. Among the firms which produced LED watches there was a keenness to embrace the technology, but developing the necessary expertise to exploit it in the end proved too much for them. In Schumpeterian terms, the turbulence in the US market was not a 'gale of creative destruction', but one of 'total destruction'. Compared to 1975, when there were estimated to have been 40 firms in the industry, the number fell to only three (Timex, Texas Instruments and Fairchild) within a space of a few years.[33]

The 'mystery' of why the US focused on LED technology and Japan on LCD technology is relatively straightforward. Around the time when the calculator was undergoing development, or perhaps even before then, the US developed a capability to produce LEDs[34] for use in electronic calculators, but Japan did not.[35] From the standpoint of the US firm, the LED was the 'right' technology to focus on because that was the signal coming from the consumer end of the market, and also because it was highly profitable compared to LCD technology.[36] For Japanese firms though, LCD technology was the 'natural' technology to focus on. Firstly, it would appear that it had no major capability in LED technology like the US had, and secondly, a large number of articles and reports had been produced in the US and Japan in the early 1970s, stating the advantages of LCD technology, and its potential even though no one was precisely sure what form it would take.

The LCD's main advantages over the LED are its very low voltage requirements and ultra-low power consumption. One additional, but nonetheless very important feature of the LCD is that it is highly compatible with CMOS technology, which also shares the aforementioned properties. An early sign of this compatibility between CMOS technology and the LCD can be seen in Figure 7.4. From the list of twelve firms in 1972, it can be seen that eleven of the firms chose CMOS technology as the circuitry for their watches. Out of these only one firm chose the LED as the display, while three firms opted for the LCD as the display. Needless to say, the LCD's compatibility with CMOS technology has been one of the principal factors, which has helped propel the LCD's enormous growth over the past two decades.[37]

The evidence suggests the majority of US firms only adopted LED technology as a 'stop gap' measure which would give them a temporary first-mover advantage in the digital watch industry. From what can be deduced of what was going on in the industry at the time, the intention of US firms was always to move over to LCD technology, but only when the reliability problems with LCD technology had been sorted out. One of the critical problems, which dogged the very early LCDs, was the stability of the liquid crystal. This problem was resolved when a chemical, cyanobiphenyl, developed by Professor George Gray at Hull University, was introduced onto the market in 1973.[38] Whilst it was the intention of many US firms to switch over to LCD technology at the 'appropriate' time, when it came down to it, they quickly discovered they were way out of their depth. A major cause of their failure was that they had become 'infected' with the

Firm	Circuit	Display
Benson (US)	C/MOS	Conventional
Bulova (US)	C/MOS	Conventional
General Time (US)	C/MOS	Conventional and digital
Hamilton (US)	C/MOS	LED
Microma Universal (US)	C/MOS	Liquid crystal
Timex (US)	C/MOS	Conventional
Ebaches/Longines (Switzerland)	Bipolar	Liquid crystal
Girard Perregaux (Switzerland)	C/MOS	Conventional
Omega (Switzerland)	C/MOS	Liquid crystal
Waltham (Switzerland)	C/MOS	Liquid crystal
Smith's Industries Ltd (UK)	C/MOS	Conventional
Seiko (Japan)	C/MOS	Conventional

Figure 7.4 What's what in electronic watches
Source: Table adapted from 'A case for new watchmakers', *Electronics International*, 22 May 1972, p. 60

Hayes & Abernathy syndrome. Rather than concentrating on making preparations for the changeover, they allowed themselves to become blinded by the continued profitability of LED technology.

The two reports quoted below give an idea of how extensive the optimism about LCD technology was during the early 1970s. The first comes from the Japanese engineering journal, *Nikkei Electronics* (1972). It stated:

According to the forecast in the report of the Investigation Committee on New Functional Devices in Foreign Countries, which was dispatched by the Japanese Association for the Development of Electronic Industries, the market for LEDs will be saturated by 1975 because of the growth of LCDs. The following viewpoint is widely shared. 'We cannot predict what will become of LCDs eventually, but they will start to be utilised in portable products, which will lay the foundation for developing newer technologies and further demands'.[39]

The second comes from *The Electronic Watch: Its Time Has Come*, published by the US Quantum Science Corporation in 1973. It reported that:

LCDs are technologically and economically doomed to be dominant. The manufacturing problems of liquid crystals ought to have been solved in 1972 and the price per digit will be half of LEDs. Because they consume less power, they will gain the major position in the digital watch market.[40]

The LCD watch finally 'triumphed' over the LED watch because by 1977 its principal weaknesses had been rectified and its overall performance exceeded the LED's. The LCD watch could operate for more than a year with a small cell battery; it could function within a temperature range of between –10°C and 60°C; and the display had a life span of more than five years.[41]

The very high failure rate of US firms to make a successful transition to the LCD watch can be largely attributed to their chronic weakness in the production field. The extent of the problem was so bad that Castellano (1988) notes that from 1976 onwards, it was not high-quality US-made LCD watches (and calculators) that became common household items, but Japanese-made ones.[42] An example of the type of technical problem which plagued US firms, which incidentally they failed to overcome quickly, was the extremely slow speed of the LCD to multiplex (the simultaneous transmission of several messages along a single channel of communication). As far as they knew the only way round this was to insert a specific component into the watch, but this caused more problems than it solved. It pushed up production costs and the failure rate of watches.[43] The other major problem, whose importance they underestimated, was the automation of the production lines.

Automation

These production problems were not confined to the small or medium-sized firms, they proved very troublesome for the large firms still left in the digital watch industry after the shakeout. According to the manager of the Time Products Division of Texas Instruments in 1977, production costs would remain high until the LCD production lines had been as extensively automated as the LED production lines. Up to that point, he estimated that only half of the LCD production lines had been automated.[44] The LED production lines had been extensively automated to insure high utilisation rates to lower costs.[45] The automation of the LCD production lines seems to have been constrained by worries about falling capacity utilisation rates and the knock-on effects this would have on profitability. Texas Instruments could not see the advantage of increasing its production of LCDs when it saw Beckman Instruments, which was the country's largest LCD producer, operating at 70–80 per cent of capacity after it had been operating at full capacity seven days a week, and Motorola considered current LCD production to have been ample.[46]

Timex was another firm which did not find the going very easy. It took the firm three attempts before it was able to make a successful entry into the digital watch industry, and this was only facilitated after it had purchased RCA's LCD facility and organised an electronics division.[47]

The strategies adopted by Texas Instruments and Timex indicated that they were intent on staying in the digital watch industry, and saw opportunities to be exploited. In 1977 both firms went on the offensive, Texas Instruments startled the digital watch industry by unveiling a range of 15 LCD watches priced between $27.95 and $48.95, and Timex introduced into its Marathon line, 10 LCD watches, some with six digits and a stopwatch priced between $46.95 and $65. This increased the firm's range of LCD watches to 41.[48] Unlike its competitors, Fairchild's strategy was to focus on promoting its brand name and improving its distribution network. After 1977 the conditions in the digital watch industry remained as competitive as ever. It was not the competition between Texas Instruments, Timex and Fairchild which kept the market vibrant, but the increasing presence of Japanese firms which had filled the vacuum created by the collapse of the US competition.

The Japanese presence completed the process which the previous shakeout had failed to do, the near total annihilation of every US firm in the digital watch industry. Fairchild was forced to leave the industry in 1979, and was later followed by Texas Instruments in 1981,[49] which left Timex as the only US firm still in the industry. The 'disappearance' of US firms from the digital watch industry had a very substantial knock-on effect on the domestic LCD industry.

Throughout the 1970s there was an array of LCD firms, most of whom produced LCDs for the watch industry on one basis or another. Some of the large vertically-integrated semiconductor firms produced LCDs for in-house purposes, while other firms supplied LCDs to the watch firms directly. This meant that the fortunes of many LCD firms were inextricably linked to those of the watch firms. In other words, when the watch firms caught a fatal dose of flu so too did the LCD firms. Fairchild epitomised the whole situation. In 1975 it bought a firm called Princeton Materials Science, which made liquid crystal materials and displays. This acquisition made Fairchild the country's second largest LCD producer, and strengthened its overall position.[50] After it was forced out of the digital watch industry, it promptly ceased the production of LCD watches.[51] Beckman Instruments was not immune either. By 1981 the situation had deteriorated so badly that it too was forced to cease production of LCD watches, because by this point the rest of the US LCD industry had become virtually extinct.[52]

The conclusion one must draw from this is that to US firms the 'only' use for the LCD was in the digital watch. Once they had lost the digital watch market, which rendered their capability in LEDs 'useless' at a stroke, the firms were at a loss about what to do next. Their lack of a proper market concept for display technologies, meant that for those firms determined to survive in the LCD industry their only option was to seek out 'niche markets'.

The mass exodus of firms from the industry is best exemplified by Motorola, a large semiconductor firm, now better known for its very strong presence in mobile phones. As part of a corporate earnings statement in 1979, the firm announced its intention to withdraw from the LCD business at a cost of $7.9 million.[53] This came as a major surprise to the industry, given that the firm had earlier devoted resources and made considerable efforts to develop its LCD operations. Both the collapse in the price of watches and competition from Japan were cited as reasons for the withdrawal.[54] Presumably, the effects of Japanese automation must have been quite far reaching. The question of utmost importance to Motorola was future profitability. The prevailing conditions in the industry suggested to the firm that it would not be able to generate 'sufficient' profits, even with a reasonable level of investment.

In the end, Motorola decided it would be much easier to generate a steady stream of profits from its core business, than if it tried to maintain a presence in an industry it did not know as well and attempted to compete with firms which were well established. In watch-display modules alone, the price per unit declined by 50 per cent in a year, to well below $1 each. The collapse in price was blamed on Japanese firms, especially Sharp, which had recently brought on stream a $10 million automated production facility.[55]

The speed with which firms were leaving the industry prompted the question as to whether Motorola would even be able to find a buyer for its LCD business; this was thought to be unlikely unless it was sold as a turnkey package.[56] By the time the major rush of firms had passed, all that remained of the industry were a few small producers who concentrated on niche markets, where they faced no competition from Japanese firms. What is interesting to note here is that the initial rush of firms into the industry and their subsequent exit is very typical of US firms' pattern of entry and exit in industries once profit margins have come under severe pressure or have totally evaporated.

The Japanese threat to the US in LCDs was recognised in 1970. In an article by Joseph Castellano, who was one of the original members of RCA's LCD research team during the 1960s, it was noticed that Japanese firms displayed the greatest resolve of any of the US competitors to exploit LCDs.[57] As Castellano states, 'Japan also boasts extensive liquid crystal development programmes. But the Japanese frankly admit they're still chasing RCA's results'.[58] The most active US firm in the development of liquid crystals in 1970 was RCA, and as mentioned in Chapter 6, it had helped put the US in a prime position to exploit the technology. This explains why Japanese firms were so eager to replicate RCA's results. Once this technological gap had been closed with the leading firm in the US, it would mean the serious problem that had confronted them in semiconductors would have been eradicated once and for all.

As far as the digital watch industry was concerned, the leading Japanese firm was Seiko. The technological capability of Seiko at the beginning of the 1970s would indicate that the technological gap, if there ever really was one, between the US and Japan was in fact very small. Seiko started work on liquid crystals in late 1968, by 1970 it had already built a prototype LCD for a clock, and by the middle of 1971 it planned to launch a model of the clock on to the mass market.[59] It soon became clear this was the technology the firm intended to develop in the future.

The principal weakness of LED technology, the very high power consumption, was a major constraint on product development. When Seiko launched its Pulsar watch with a heavy gold case in 1969, sold only in Japan for $1,250, the weakness of LED technology was all too evident. When the user pushed the on-command button to show the time on this luxury watch, the very high-power consumption only permitted the hour and minute light to be displayed for 1.4 seconds![60] Seiko's development of an LED watch, and Ricoh's decision to opt for LEDs for use in its digital watches, confirms that Japan had been developing LEDs and LCDs simultaneously, just like the US, before it decided to 'target' LCDs.

When Seiko started work on liquid crystals in late 1968, it was not the only Japanese firm to show an interest in the technology. Other very well known firms such as Hitachi and Sharp also began work on liquid crystals that put them in a good position when the tide turned away from LEDs and towards LCDs. What they did not know was that when the tide turned in their favour, they would be left with an industry devoid of any significant overseas competition. As mentioned previously, it was the LCD watch which came to dominate the Japanese digital watch industry even though a firm such as Seiko had the capability to develop an LED watch. Apart from the Japanese consumer showing a preference for LCDs earlier than the American consumer, there were two other features about the Japanese watch industry that made it very distinct from the US industry.

Firstly, as noted earlier, firms did not swarm into the industry as the watch market grew and matured. Seiko was one of only four firms which had traditionally dominated the conventional watch industry, the others were Citizen, Ricoh and Orient. Throughout the 1970s, the only other firm prepared to enter the Japanese industry and compete with the 'Big Four' was Casio, a vertically integrated calculator producer.[61] Casio was able to make a highly successful entry because it already had a very well established distribution network, and chose to challenge the traditional firms via the normal channel, the jewellers. According to the managing director of Casio, success was not immediate or easy, it took some three years before it was accepted by the industry and jewellers alike.[62] Its entry was further facilitated by its technological competence in LCDs gained from its experience in the production of calculators.

The second feature was the absence of any shakeout among the firms in the watch market at the time when there was a blood bath taking place in the US market. Any shakeout failed to materialise primarily because there was no 'switch' from one technology to another. As shown in Figure 7.5, the demand for LCD watches increased between 1975 and 1978 at a pace which was favourable to all the firms. The gradual rise in demand without any swarming of firms into the industry ensured prices remained stable; the watches at the lowest end of the market retailed for around $50. With prices at this level no firm was placed under any financial pressures as the market could comfortably accommodate all those firms present. Rather than having to face a financial 'black hole' which confronted the US firms, Japanese firms were instead looking forward to a period of steady growth.[63]

The methods used by the Japanese watch firms to obtain their supplies of LCDs was identical to those used by many of the US watch firms. Seiko and Citizen chose to become in-house producers and started production of TN-LCDs in 1973, although prior to this date Citizen had obtained some of its supplies from Hitachi. On the other hand, from 1974 Orient depended on Sharp for all its supplies of LCDs, and Casio increased its dependence on Hitachi from the mid-1970s to the late 1970s, although from 1980 onwards it started to reverse this process by becoming more self-sufficient and producing LCDs in-house.[64] Here is where the resemblance between the Japanese and US watch industries ends.

Unlike the US watch firms, the Japanese watch firms were very careful, so to speak, not to put all their eggs in one basket. As well as supplying the domestic market, they were also highly export-oriented, with their main export market being the US. This gave them both considerable protection from any serious downturn in their domestic market and offered them an opportunity to see how they compared with the best US firms. Figures cited by Numagami (1996) illustrates how important exports were to the Japanese firms, and the scale of their subsequent success after 1977 when the collapse of the US firms occurred. In 1978 Japan exported 23 million units, 47 per cent of the total production of 48 million units. Compare this to the US, which only managed to export 6 million units (of all types), 20 per cent of a total production of 29 million units – and this was in the year when its watch exports reached a peak.[65] One important point that cannot be overlooked is that until 1976 most Japanese exports comprised of watches with a conventional face and hands. Nevertheless the surge in Japanese production between 1977 and 1985 compared with the US, as illustrated in Figure 7.5 is quite astonishing.

The other major distinction between the US and Japanese watch industries was the very different strategies pursued by the LCD

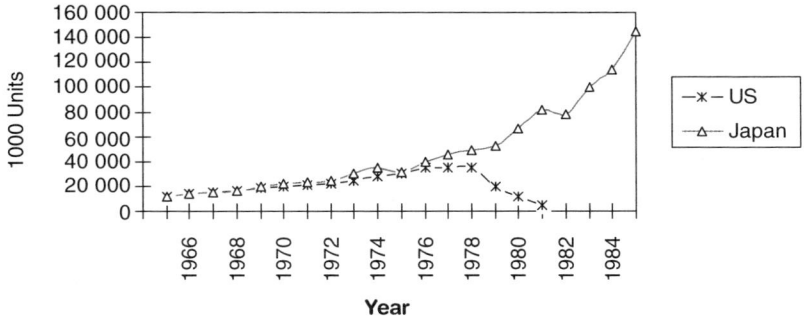

Figure 7.5 Production of watches by the US and Japan between 1965–85
Source: Graph adapted from Figure 5 in Numagani, 1996:151

firms. For Beckman Instruments, which was the leading producer in the US, the digital watch industry was one to which it could supply a standard mass-produced product. The firm, like other US firms producing LCDs, appears to have been very reluctant to invest in new product and process innovations, one of the visible signs of a firm suffering from the Hayes & Abernathy syndrome. Contrast the strategy of Beckman Instruments with that of Sharp and the differences could not be greater. Sharp, which has been the leading Japanese firm in the industry since the early 1970s had the same strategy as the Japanese watch firms, in so far as it was careful not to become too reliant on a single source of demand. Not only did it foster long-term relationships simultaneously with firms producing watches and calculators; it was also very active in the calculator business itself, which saw it launch the first calculator with a TN-LCD in 1973. This provided it with a good foundation on which to build its LCD business, as it provided it with the necessary earnings required to help finance the enormous costs of R&D, and gain valuable production experience of manufacturing LCDs for different applications.

To help overcome the problem of costs, when Sharp started research on liquid crystals in 1970, it did not try and do everything on its own. Rather it persuaded other firms to get on board the LCD bandwagon which was beginning to gather momentum, and help it develop liquid crystal materials and equipment necessary for LCD production. One firm with which Sharp developed a highly successful partnership is Dainippon Ink Chemicals (DIC). It learned how to evaluate the electro-optic properties of the organic materials, and by the mid-1970s it had started to produce equipment for LCD production.[66]

With the income generated from its involvement in both the watch and calculator industries, combined with its partnerships with supplier firms like DIC, Sharp led the development of the second generation of LCDs, known as the super-twisted nematic (STN) LCD. This is an advanced version of the TN-LCD. The feature which distinguishes the STN-LCD from the TN-LCD, is that the 'twist' of the liquid crystal molecule is 270° in the STN-LCD, compared to only 90° in the TN-LCD. The 'extra' twist greatly enhanced the performance of the LCD, giving it a better contrast, an improved viewing angle (up to 40°), and making the use of colour possible. The STN-LCD was put into mass production in 1986. It might well have appeared on the market sooner had it not been for a major problem with the original version. What prevented it from being launched any earlier was that when light was passed through the liquid crystal there was a substantial amount of refraction. Because the light was sent in a number of different directions (caused by the reaction of the molecules), a user looking at an STN-LCD would see black text on a yellow background or a white text on a blue background. The problem was eventually solved when it was found that when another layer of liquid crystal material was placed on top of the first layer, the situation corrected itself. As light passes through the second layer it twists for a second time, and as it does so it cancels out the refraction caused by the first layer of liquid crystal.

Although the development of the STN-LCD opened the way for the LCD to be used in applications other than watches and calculators, the overall performance of the STN-LCD was still inferior to the traditional cathode-ray tube (CRT). Amongst other things, the CRT's picture quality, viewing angle, and contrast was still much better than anything the STN-LCD could manage. Research on the STN-LCD continued apace to improve its overall performance. Considerable progress was made, with the launch of the double super-twisted nematic (DSTN) LCD in 1987, and the triple super-twisted nematic (TSTN) LCD in 1989 (all of which were patented by firms such as Sharp and Hitachi). Even with these improvements, the performance of the LCD was still not on a par with the CRT. This had to be resolved if it was ever going to make inroads into the CRT's traditional markets such as monitors and televisions.

The relatively poor performance of the STN-LCD vis-à-vis the CRT was primarily due to the way it is fabricated. Like its predecessor, the TN-LCD, the STN-LCD is a passive display. This type of fabrication has serious technological limitations, the contrast is still insufficient to reproduce the full range of colours available with a CRT, and the response times are relatively slow which mean that moving images tend to blur and leave shadows.[67] Before the LCD could start to challenge the CRT, its performance would have to undergo a quantum leap.

At the same time as Sharp had been developing the STN-LCD, it had also been doing research on TFT technology, originally developed by Peter Brody at Westinghouse during the 1960s and early 1970s, and quickly emerged as the industry's leading firm. As early as 1972 the Japanese engineering journal, *Nikkei Electronics*, reported that research on TFT technology was advancing.[68] By all accounts, Japanese firms were now convinced that TFT technology was the technology that could finally unleash the LCD's potential to challenge the CRT.

The creation of the technological gap

Put into perspective, the creation of the technological gap which Japan opened up with the US is not that surprising. One has to look no further than the semiconductor industry to appreciate that the failure of the US watch firms to make a successful switch from LED to LCD technology is far from being an isolated case. Foster (1988) points to the fact that out of a total of ten US vacuum-tube manufacturers, none of them managed to make a successful transition to the use of transistor technology when the changeover occurred in the mid-1950s.[69] Six of the firms failed to significantly appreciate the revolutionary nature of the emerging technology, and the other four, which did take the threat of the new technology seriously, tried to make the transition, but failed. In the end none of them became major semiconductor producers.[70]

Foster goes on to say that the inability of firms to make a successful switch from the use of one technology to another is not only confined to high-technology industries. A good case in point is that of tyre cords. The pattern that emerges from this example is that one technology (or material) is 'dominant' for a certain time period, only to be then substituted by another. With each substitution, there was a change of leadership in the industry. Between the 1920s and mid-1940s, tyre cords were made of cotton. Then in the mid-1940s, cotton was replaced by rayon, and the American Viscose Company became the industry's leading firm. Rayon was subsequently replaced by nylon in the early 1960s, and the American Viscose Company was displaced by Du Pont. Finally, when nylon was replaced by polyester fibres in the late 1970s, Du Pont lost its top position to a firm called Celanese.[71]

When firms in industries as diverse as digital watches, semiconductors and tyres, stretching over a number of decades, fail to make a successful switch from one technology to another, it suggests there are common factors at work which eventually leads to a firm's downfall. One factor that undoubtedly contributed to the general collapse of the US digital watch industry was the technological limitation of LED technology, especially its high power consumption. As there was no quick or obvious solution to this problem it was only a matter of time before it was superseded by a technology known to be much better.

The other critical factor which plagues any firm heavily committed to an 'old' technology and threatened by an emerging one, is determining the optimal time to make any switch. Any firm committed to an old technology stands to lose a considerable amount of profits if it decides to make the change too early. Even though the US firms knew in the early 1970s that the chances of LED technology being superseded by LCD technology were very high, the majority of them do not appear to have made the changeover an urgent priority while the demand for the LED watch remained strong and there were easy profits to be made. By allowing profitability to get the better of them, the firms' 'technology strategy' went out of the window.

A more fundamental problem for firms is when should they start their preparations to deal with the consequences of an emerging technology? It is all very well acknowledging a technological threat, but actually taking steps to prepare for it is another matter. No firm can ever be sure exactly when the threat of a new technology will begin to have a major impact. The time factor is something of a moving target, as it could be anything between, for example, one and ten years. The example of the US watch firms is an excellent illustration of a group of firms which were aware of the threat of LCD technology, but when the time actually came to make the switch, they were so ill-prepared only three firms survived after 1977.

One of the reasons why the threat of LCD technology was completely miscalculated by the US watch firms is because they failed to realise how long it takes to develop a new technology and the expertise to exploit it. Foster claims the length of time required can be anywhere between five and ten years.[72] As an estimate, this seems very realistic in view of the fact that there are a nine-year and a seven-year gap respectively, between the dates Seiko and Sharp started research on liquid crystals and the surge in LCD watch production after 1977. Although LCD watches were produced in significant quantities prior to this date, the success of the Japanese firms in the US market would indicate that the reliability and quality of their LCD watches were far better than the first ones to appear on the market.

Firms like Seiko and Sharp had clearly gained a valuable under-standing of the technology, and were in the position to exploit it. If Japanese firms say took an average of eight years to obtain an under-standing of the technology, those US firms which attempted to make the switch did not have a chance given the very short time frame they had to work within. It is hardly surprising that the US firms experi-enced major difficulties when they had given themselves little more than a year or two to get to grips with something which had taken the Japanese industry several years to learn how to master. Furthermore, the US watch firms did not have the luxury of being able to fall back on the domestic LCD industry to supply them with LCDs that were of a comparable quality with those produced by Japanese firms. In the available literature, there are no reports of any US LCD firm having a major research programme similar to those which were commonplace in the Japanese industry.

The other error made by the US watch firms was misinterpreting the market.[73] In a country where the concept of the 'market' is held in very high esteem, it is all too easy for a firm to become blinded by it. The market inadvertently sustained the watch firms' commitment to LED technology because of the relatively high level of demand. On the surface, the consumer appeared satisfied with the LED watch and this was sufficient to convince the firms that they were on the right tracks. When profits came under pressure, this could easily have been inter-preted as a sign that the industry was beginning to mature and the process of consolidation was on the verge of getting under way. The firms' judgement was right in so far as consolidation was concerned. There are always going to be casualties in an industry as it approaches maturity, especially one with so many firms all scrambling for a share of a limited market. Where their judgement was completely out was in the speed with which consumer tastes can change.

They were caught out not just by the switch to another technology, but also by the American consumer's new found appetite for Japanese watches. Even those US firms that sold LCD watches missed out; Japanese watches had gained a reputation for quality and reliability, something which was not equated with watches of US origin. The

American consumer's faith in US-made watches had been badly shaken by their experience of the LED watch, which were described in one report as being 'shoddy'.[74] Apparently, US industry executives were worried that unless the issue of quality was sorted out by firms, and they produced LCDs to a much higher standard than they had done with LEDs, the market for digital watches would be seriously affected.[75] Japanese industrial executives, on the other hand, had no such worries. Japanese firms appear to have sorted out any lingering problems with the quality of their LCDs in the early 1970s prior to them being put on the market. This is backed up by Yoshio Yamazaki of Seiko-Epson who said consumers reported no reliability problems stemming from the TN-LCD in their first watch which the firm introduced in 1973.[76]

The effects of the upheavals in the digital watch industry continued to be felt well into the 1980s. If anything, the upheavals helped precipitate a widening of the technological gap which had already begun to appear in the 1970s. As mentioned earlier, all that remained of the US LCD industry after 1981 was a handful of firms that focused on niche markets where Japanese competition was absent. Under these circumstances, after an industry of this kind had undergone such a huge contraction, one might have assumed that it would have gone through a process of consolidation, and then tried to rebuild itself. In fact, what happened was quite the reverse, the process of contraction continued throughout the 1980s, as shown in Figure 7.6. After 1980 a further seven LCD operations were either sold or closed down, which left the US without a single production facility capable of producing LCDs on a commercial basis. For all intents and purposes, this indicated the industry had finally come to the end of the road.

As the US industry continued to disintegrate, the Japanese industry found itself presented with an opportunity almost too good to be true. Here was an industry with enormous potential, yet strangely enough it was completely free of any major competitors, indicating beyond any reasonable doubt that Japan had a radically different concept of the market for LCDs than any other country. Without any competition to worry about, the Japanese firms started to build upon their early advantage. Initially though, the build-up was slow because the firms found themselves caught up in a kind of technological trap. The firms knew that the technological limitations of the STN-LCD was hindering the demand for LCDs, but paradoxically, they could not boost it even though they had the technology which could in their possession. The TFT-LCD still required an enormous amount of development before it could rival the performance of the CRT.

Firm	Year
Hewlett Packard	Closed 1980
Texas Instruments	Closed 1980
Kylex/Exxon	Sold 1983
Crystal Vision	Closed 1984
Panelvision	Sold 1986
RCA	Sold 1987
Honeywell	Closed 1988
LC Systems	Closed 1988
General Electric	Sold 1989

Figure 7.6   US LCD operations sold or closed in the 1980s

Source: Table adapted from Borrus, M & Hart, JA (1994) 'Display's the Thing: The Real Stakes in the Conflict over High-Resolution Displays', *Journal of Policy Analysis and Management*, Vol. 13, No. 1, p. 47

To get out of this trap the firms adopted a dual strategy. They continued research on STN-LCDs to improve their performance, but at the same time they stepped up their research efforts to develop the TFT-LCD. As early as 1983 it had become clear that Japanese firms were determined to exploit the TFT-LCD and overcome the development problems. In an article in *Electronics International*, aptly titled 'Japanese flat panels flatten competition', it was noted that Japanese firms were accelerating their production of LCDs, and that US firms were withdrawing from the field of flat panel displays generally.[77] The additional output of LCDs found their way into the then new consumer electronics like hand-held TVs, portable computers, and car dashboards.[78] The article also mentioned how important the Japanese firms' experience in calculators and digital watches had been to them. Through their involvement in these two industries, and the production of 20 million five to seven character LCDs per month, they had refined their production techniques sufficiently to achieve high-yield, low-cost processes to produce the newer, larger colour displays.[79] By 1986 this very high level of efficiency in production had started to yield some very positive results. The Japanese firms had developed a number of different sized TFT-LCDs, for example, Seiko had developed a 14-inch TFT-LCD, Matsushita a 12.5-inch TFT-LCD, Mitsubishi a 10-inch TFT-LCD, Hosiden a 7.2-inch TFT-LCD, and Hitachi a 6.3-inch TFT-LCD.[80]

The role of MITI in all of this had been minimal. In the early 1980s, it instigated a number of initiatives such as the establishment of the Large Circuit Element Technology Development Institute together with firms such as NEC, Sharp, Casio and Sanyo to develop large-scale colour LCD devices for 'wall hanging' televisions and ultra-thin copiers.[81] In addition, MITI organised at least two consortia, the Giant Electronic Technology Corporation, involving 17 firms, to develop LCDs for high-definition television (HDTV) applications, and the HDTech consortium to develop LCDs for use in HDTV.[82] Apart from this, the Japanese firms have received little other support.[83]

Unlike the STN-LCD, which is cheap and easy to produce, the TFT-LCD is phenomenally complex and costly to build. For example, to produce a 10-inch TFT-LCD takes over 100 manufacturing steps spread over a four-week period. The display consists of between 10–15 layers of material and involves etching around 4 million components. Furthermore, one faulty transistor or a single dust particle is sufficient to wreck a complete display,[84] so the production facilities have to be very highly automated and extremely clean. For a firm contemplating the production of TFT-LCDs, this level of automation is a form of entry barrier. Unless it can at least match this standard of technical know-how and expertise, any foray it might make into the industry is doomed to failure. The exacting production requirements to make a TFT-LCD has meant the limits of cleanliness have reached new peaks. At Sharp's Tenri plant, the air inside is one million times cleaner than the air outside.[85] Constant maintenance of an environment of this kind, in addition to the need for precision engineering of the very highest order, have meant that production yields have been slow to rise. Around 1992 it was estimated that yields of most of the major manufacturers were approximately 50 per cent, and by 1994 they had increased to approximately 80 per cent.[86] The question of yields remains extremely sensitive within the industry as from that information it is possible to work out how much profit is made, or for that matter, the loss incurred on each display.

The complexity of the TFT-LCD posed enormous challenges even for Sharp, which is at the cutting edge of the technology. The problems start as the size of the display increases, so to be able to develop the

TFT-LCD, the only option open to firms has been to produce small displays, and then gradually increase the size of the display as their understanding of the technology improves. This incremental approach has considerably slowed down progress towards the wall hanging TV, the ultimate goal of George Heilmeier and his team at RCA in the late 1960s.

The major benefit that Japan derived from the turmoil in the digital watch industry was that it severely weakened a major potential competitor in LCDs. The other element of the digital watch industry episode that also worked very much to Japan's advantage, and was quite fortuitous, was that the Japanese consumer rejected the LED watch from the outset. Quite inadvertently, this eliminated at a stroke any chances of the same kind of turmoil which hit the US industry from hitting the Japanese industry. With US LCD firms in a parlous financial state from the mid to late 1970s, and a stable domestic digital watch industry, Japanese LCD firms have been fortunate enough to develop LCD technology almost completely unhindered. The strategy which Japanese LCD firms have followed of making incremental improvements to LCD technology has insured that their progress in the industry had been built on very solid foundations. With such extensive expertise in LCD technology, it comes as no surprise that by the early 1990s Japanese firms had gained a stranglehold on the TFT-LCD industry.

With hindsight, from where the Japanese firms stood in the early 1990s, they were acutely aware that the growth of the industry would attract competitors. The unanswered question then was where the biggest threat to their dominance lay. Based on previous experience, the Japanese firms had much to fear from their South Korean competitors, and perhaps those from Taiwan. As the 1990s progressed, what the Japanese firms would have had absolutely no inkling of would be the speed at which the competition would break their stranglehold and then leave them floundering by the decade's end. Japan's global share has declined at a surprising speed. In 2001 Japan's global market share stood at 50 per cent,[87] but by 2003 it had declined to 34 per cent,[88] and was expected to decline even further to 15 per cent by 2006.[89] This rapid loss of market share has forced the Japanese firms to consolidate and regroup, just as they have done in semiconductors. Two of the most recent developments are Sony's joint venture with Samsung, known as S-LCD, which commenced operations in 2004,[90] and the joint venture formed between Hitachi, Toshiba and Matsushita in the same year to produce TFT-LCDs for televisions.[91]

Summary

From the 1970s onwards the strategy of Japanese firms is very clear. They were determined to develop the LCD over the long term. In the path that they set out for themselves, they were lucky, as mentioned above, that any potential competition from the US was dealt a fatal blow with the turmoil in the digital watch industry. There is no question about the capability of Japan's NSI to support the development of LCD technology or the capacity of the firms to develop it. What is the most surprising element here is how the Japanese firms failed to protect and upgrade their competencies in TFT-LCDs as it became more obvious that firms from South Korea could compete with them on more or less equal terms. Very quickly, the Japanese firms found themselves in the most unusual position of playing catch-up in a technology where they had not too long ago enjoyed unrivalled dominance.

Notes

1. The sources which are referred to are: Brody, TP (1984) 'The Thin Film Transistor: A Late Flowering Bloom', *IEEE Transactions on Electron Devices*, Vol. ED-31, No. 11, pp. 1614–1628, Brody, TP (1996) 'The birth and early childhood of active matrix: A personal memoir', *Journal for the Society of Information Display*, Vol. 4, No. 3, pp. 113–127; Florida, R & Browdy, D (1991) 'The Invention the Got Away', *Technology Review*, Vol. 94, Part 6, pp. 43–54; Tyson, L (1992) *Who's Bashing Whom: Trade Conflict in High-Technology Industries*, Washington, DC: International Institute for Economics.
2. Tyson, 1992:18.
3. *Ibid*. The Organisation for Economic Cooperation and Development (OECD) uses the International Standard Industrial Classification (ISIC) to identify high-technology industries. Under this system the OECD has classified: drugs and medicines (SIC 3522), office machinery and computers (ISIC 383, but excluding 3832), electronic components, aerospace (ISIC 3845), and scientific instruments (ISIC 385) as high technology industries (*Ibid*, Table 2.1:21). Under the Guerrieri-Milana classification system which is based on the OECD's own definition, the following have been classified as high technology industries: chemicals (synthetic organic colouring, products for agriculture, radioactive materials, polymers and plastics), pharmaceuticals (antibiotics and selected other products), power generating machinery (turbines and piston engines), electrical (power machinery and selected apparatus), data processing (machines, processing and storage units and parts), electronic office machines (photocopying apparatus, other), telecommunications (telephone, telegraphy, and transmission apparatus and, selected equipment, parts and accessories), electronic components (integrated circuits and microassemblies, semiconductors, television picture tubes, selected parts), aircraft (aircraft, helicopters, spacecraft, reaction engines), scientific instruments

(electronic measuring and controlling instruments, particle accelerators, optical instruments) (*Ibid*:21–22). The source for the Guerrieri-Milana classification which Tyson refers to is Guerrieri, P & Milana, C (1991) 'Technological and Trade Competition in High-Tech Products', *BRIE Working Papers 54*, Berkeley: University of California.

4. *Ibid*:23. Figures obtained from Table 2.3.
5. *Ibid*.
6. *Ibid*:19.
7. Here Western Europe constitutes the nine countries of the old European Community (EC). The table has been adapted from Table 2.5 to illustrate the point that Western Europe and the US have conspicuously failed to combat the continued weaknesses of their respective electronic industries, which can be broadly measured by the loss of market share.
8. Sharp, M (1993) 'The Community and new technologies' in J Lodge (ed.) *The European Community and the Challenge of the Future*, London: Pinter, 2nd edition, pp. 201–203.
9. Patel, P & Pavitt, K (1987) 'Is Western Europe losing the technological race'?, *Research Policy*, Vol. 16, Nos. 2–4, p. 60.
10. Numagami, T (1996) 'Flexibility trap: a case analysis of US and Japanese technological choice in the digital watch industry', *Research Policy*, Vol. 25, No. 1, p. 149.
11. Bahadur, B (1983) 'A Brief Review of History, Present Status, Developments and Market Overview of Liquid Crystal Displays', *Molecular Crystals & Liquid Crystals*, Vol. 99, p. 368.
12. 'US firms gird for computer battle', *Electronics International*, 23 November 1970, pp. 83–86. The number of Japanese firms producing calculators outnumbered US firms by a considerable margin. In what became an increasingly important sector of the electronics industry, the US had fewer than ten firms while Japan had more than twenty. With so many firms producing calculators, Japan's production rate and exports soared very rapidly. In 1969 Japanese shipments totalled 441,000 units, but by the first half of 1970, shipments had already reached 519,000 units, more than the entire production of the preceding year (*Ibid*). In the following three years production reached several million. A large percentage of this increased output was exported, especially to the US. Because of its growing success Japan attempted to slow down the rate of growth in exports fearing that this would stir up protectionist tendencies in the country. Despite MITI's efforts to 'regulate' the market it was unsuccessful ('Japan seeks curb on exports to US', International newsletter, *Electronics International*, 5 July 1971). By the first half of 1973, Japan had exported 2.35 million units which was 2.4 times the number of units exported in the first half of 1972. About half of these calculators, 1.18 million went direct to the US ('Japan accelerates calculator exports', Electronics newsletter, *Electronics International*, 16 August 1973, pp. 29–30).
13. Scherer, FM (1992) *International High-Technology Competition*, Cambridge, Massachusetts: Harvard University Press, p. 65.
14. *Ibid*:66. Because of the large number of semiconductor firms which had entered the calculator industry, a number of calculator firms decided on a policy of backward integration as they had become extremely concerned

about the reliability of supplies of semiconductors. They were under the impression that if they manufactured their own semiconductors for use in-house this would put their minds to rest. Backward integration was also popular amongst firms as a means of securing them supplies of other valuable components such as modules (Valery, N, 'Coming of age in the calculator market', Calculator Supplement, *New Scientist*, 13 November 1975, p. ii).

15. Scherer, 1992:66.
16. *Ibid*.
17. *Ibid*.
18. *Ibid*.
19. *Ibid*:67.
20. 'New IC market: electronic watches', *Electronics International*, 21 December 1970, p. 83.
21. *Ibid*.
22. *Ibid*, & 'Displays', *Electronics International*, 25 October 1973, pp. 104–106.
23. 'Displays', *Electronics International*, 25 October 1973, pp. 104–106.
24. Numagami, 1996:139.
25. 'Watch market: is 40 a crowd?', *Electronics International*, 20 February 1975, pp. 34–35.
26. 'A case for new watchmakers', 22 May 1972, pp. 59–62; 'TI defers its watch entry', 31 October 1974, pp. 29–30; 'Watch market: is 40 a crowd?, 20 February 1975, pp. 34–36; 'Fairchild seeks liquid crystal firm', Electronics newsletter, 1 May 1975; 'Timepiece uses diode, LC displays', 26 June 1975, pp. 31–32; 'Electronics adds wristwatch frills', 7 August 1975, pp. 46–47; 'Watch surge generates LCD shortage', 6 January 1977, pp. 67–69; 'LCD startup bedevils watch firms', 18 August 1977, pp. 74–75.
27. Valery, N, 'Electronics in search of temps perdu', *New Scientist*, 30 October 1975, p. 234.
28. 'Japan's watch market calm', *Electronics International*, 8 December 1977, p. 78.
29. Numagami, 1996:137, 139.
30. *Ibid*:135.
31. 'LCD display shortage is already here, says Beckman', Electronics newsletter, *Electronics International*, 9 December 1976, p. 25.
32. 'Liquid crystals for watches are turning out to be in good supply', *Electronics International*, 28 April 1977, p. 30.
33. Numagami, 1996:135.
34. Numagami states that LEDs can be traced back to 1923, but it was only during the 1960s that research was undertaken to find industrial applications for them.
35. The source for this conclusion comes from Braun & MacDonald, 1980:171. In one important sentence they write, 'The Japanese fared particularly badly for not only did they depend on MOS chip exports from the US, but they also lacked the necessary technology for producing light emitting diodes used in the calculator display'.
36. In Numagami, 1996:138, footnote 9, a French report by Comite Professionel Interregional de la Montrei, written in the mid-1970s is cited. It stated, 'According to Sears Roebuck, recent return ratio of LCD watches is virtually zero per cent, while that of LED watches is 15 per cent'.

37. *Ibid*:136.
38. Ibid:138, footnote 9.
39. Quoted in *Ibid*, 1996:140.
40. Quoted in *Ibid*.
41. *Ibid*:136.
42. Castellano, J (1988) 'Liquid Crystal Display Applications: The First Hundred Years', *Molecular Crystals & Liquid Crystals*, Vol. 65, p. 392.
43. The component was a 28-pin package for a standard 4-digit watch. 'LCD start-up bedevils watch firms', *Electronics International*, 18 August 1977, p. 74.
44. *Ibid*.
45. 'Timex goes on the offensive', *Electronics International*, 23 June 1977, p. 102.
46. 'Liquid crystals for watches are turning out to be in good supply', *Electronics International*, 28 April 1977, p. 30.
47. 'LCD start-up bedevils watch firms', *Electronics International*, 23 June 1977, p. 74.
48. *Ibid*.
49. Numagami, 1996:135.
50. 'Fairchild seeks liquid-crystal firm', Electronics newsletter, *Electronics International*, 1 May 1975, p. 25.
51. Numagami, 1996:152.
52. *Ibid*:147, 152.
53. 'Motorola bows out of LCD business', *Electronics International*, 28 February 1980, pp. 48–50.
54. *Ibid*:48.
55. *Ibid*:48.
56. *Ibid*.
57. Castellano, J, 'Now the heat is off, liquid crystals can show their colours everywhere', *Electronics International*, 6 July 1970, p. 69.
58. *Ibid*.
59. *Ibid*.
60. 'New IC market: electronic watches', *Electronics International*, 21 December 1970, p. 84.
61. 'Japan's watch market calm', *Electronics International*, 8 December 1977, p. 78.
62. *Ibid*.
63. *Ibid*.
64. Numagami, 1996:142, 145.
65. *Ibid*:139, footnote 10.
66. *Ibid*:150.
67. 'Screen Technologies: Too soon to write off the cathode ray tube', Mobile Computing Survey, *Financial Times*, 26 January 1994.
68. Nikkei Electronics, (1972) Taitosuru ekisho dispurei: Matorikusu gata no kenkyu ga susumu (Rising liquid crystal displays: Research on the active-matrix type is advancing, in Japanese), May 8, 32–43. Cited in the bibliography, Numagami (1996).
69. Foster, RN (1988) 'Timing technological transitions' in ML Tushman & WL Moore (eds) *Readings in the Management of Innovation*, New York: Harper Business, 2nd edition, p. 215.

70. *Ibid.*
71. *Ibid*:215–16.
72. *Ibid*:220.
73. *Ibid*:221.
74. 'Watch surge generates LCD shortage', *Electronics International*, 6 January 1977, p. 69.
75. *Ibid.*
76. Numagami, 1996:138, footnote 9.
77. 'Japanese flat panels flatten competition', *Electronics International*, 28 July 1983, pp. 48–49.
78. *Ibid.*
79. *Ibid.*
80. 'The Japanese are pushing work on bigger LCD panels', International newsletter, *Electronics International*, 16 October 1986, p. 50.
81. Wong, PK & Mathews, JA (1998) 'Competing in the Global Flat Panel Display Industry: Introduction', *Industry and Innovation*, Vol. 5, No. 1, p. 6.
82. *Ibid.*
83. *Ibid*:3, 7.
84. 'Screen Technologies: Too soon to write off the cathode ray tube', Mobile Computing Survey, *Financial Times*, 26 January 1994.
85. 'The future is crystal clear', *Financial Times*, 8 November 1994.
86. 'Japan's Liquid-Crystal Gold Rush', *Business Week*, 17 January 1994, p. 45.
87. 'Japanese display makers strain to regain footing', *Electronic Engineering Times*, 19 October 2001 (www.eetimes.com).
88. 'Flat out for flat screens: the battle to dominate the $29 billion market is heating up but the risk of glut is growing', *Financial Times*, 24 December 2003.
89. *Ibid.*
90. 'Joint Sony, Samsung LCD factory opens', *Electronic Engineering Times*, 19 July 2004 (www.eetimes.com). The joint venture is a seventh generation production facility located in South Korea.
91. 'Hitachi, Toshiba, Matsushita join on LCD TV panels', *Electronic Engineering Times*, 31 August 2004.

8

'Jumping Aboard' a New Sectoral System of Innovation: The Strategies of South Korea and Taiwan in Liquid Crystal Displays

In Chapter 4, it was discussed how Japan caught up with the US in semiconductors. Between the late 1980s and early 1990s South Korea turned the tables on Japan and caught up with it in semiconductors. While South Korea's success in the semiconductor industry is well known, and is analysed later, what has so far gone relatively unappreciated is the progress it has made in TFT-LCDs. Within a period of approximately five years, South Korea had become the world's second biggest producer of TFT-LCDs, and then it proceeded to overtake Japan. At the beginning of the 1990s, the joint objective of the government and chaebols[1] was to increase the country's share of the global market from 2.5 per cent in 1992 to 7.5 per cent by 1997.[2] That objective has not only been achieved, it has been surpassed by an enormous margin. In 1997 South Korea's global market share was estimated to have been anywhere between 10 per cent,[3] and 19.9 per cent.[4] By 2003 it had increased to 44 per cent.[5] South Korea has managed this by using a very similar strategy to the one used in the semiconductor industry.

As with Japan's earlier success in semiconductors, it is important that South Korea's success in TFT-LCDs is viewed from a wider perspective. In the first section of this chapter it is noted that since the early 1960s South Korea has made a conscious effort to emulate the Japanese model of economic development. With this, South Korea has demonstrated an astonishing determination to compete head-on with Japan. Semiconductors were the first high-technology industry where South Korea showed not only its determination to compete with Japan, but more importantly, its capability. The TFT-LCD industry is, of course, the second high-technology industry where South Korea has demonstrated its capability to compete with Japan. The speed with which the country has built up its presence in the industry is testimony to this determination.

In complete contrast to South Korea, which has rushed into the TFT-LCD industry, Taiwan has adopted a far more cautious approach. The approach that Taiwan has adopted has been for a good reason, the country is all too aware of the high risks that are involved in an industry of this type. The intensity of competition, the huge investment outlays, the rate of technological change, and the changing market for different sizes of displays have all contributed to it approaching the TFT-LCD industry with the caution it has. The need to get the strategy right cannot be emphasised enough. One 'wrong' move could cost Taiwan potentially hundreds of millions of dollars of investment. On the other hand, Taiwan could not afford to be too cautious because in such a dynamic industry it would have found it exceedingly difficult to establish the presence in the industry it wanted to. One of the interesting aspects about Taiwan's approach, compared to that of South Korea, is the extent to which it has been dependent on Japan for building the foundations of its own TFT-LCD industry. This not only gives an indication of the ground that Taiwan had to make up to become a global player, but also of the differences in the technological capabilities that exist between Taiwan and South Korea in TFT-LCDs. In spite of the obvious challenges faced by Taiwan, the country has shown itself highly capable of overcoming the original hurdles it faced.

South Korea: catch-up in practice

The similarity in structure between South Korea's NSI and Japan's NSI suggests that the relationship which exists between government and industry in South Korea would be of a very similar kind to that in Japan. However, this is not actually the case. Even when compared with Japan, government intervention in industry has been very extensive. The 'excessiveness' of this intervention can be almost entirely attributed to the country's backwardness and turbulent history.[6] Suffice to say, the date which can be considered to be the main watershed for the very active role in industry, which the government took upon itself, is May 1961, when the military seized power. The new military dictator, Park Chung Hee, who ruled until 1979, put the country on its path of development by making two very important decisions at the time that he took power. The first was that the country would try and emulate the Japanese model of economic development. The second, which reflected another Japanese goal, was that the pursuit of economic growth was to take precedence over everything else.

As far as planning was concerned, the only practical way for the government to steer the country along the course it wanted to follow was to set out its plans in advance in the form of a five-year development plan, and to take control over the limited amount of resources which the country possessed at the time. This way the government had complete control over important matters such as the geographical location of specific industries and could determine the planning, timing and levels of investment. These early attempts by the government to steer the economy, whilst at the same time keeping a very tight grip on industry, has created an unequal relationship between the two parties, where the government always has the upper hand.

For the government to have any realistic chance of achieving its objectives, it had to devise a way of encouraging the chaebols to diversify into industries where the risks were high as well as low. It came up with an ingenious system based on performance and incentives, whereby a successful chaebol was handsomely rewarded, and a chaebol which was judged to have been a poor performer, or was simply unsuccessful, could be severely punished. It was the old 'carrot and stick' approach used by the government, which permitted a chaebol to amass more and more carrots, the more successful it was. Failure by a chaebol meant it stood a good chance of being hit by a stick, strong enough to completely destroy it. One vital point that needs to be made here is that the South Korean government measured performance on a very different basis than is commonly used in the industrialised countries of the West. Good performance is evaluated in terms of production and operations management rather than financial indicators.[7] A chaebol's performance was measured by the progress it made in any one of three main areas, exports, R&D, and product innovation. The most problematic area for any chaebol to significantly profit from the incentive system was in exports, where the government laid down the stiffest performance targets of all.[8]

For a chaebol which posted a good performance in any one of the three areas mentioned, the 'carrots' it received from the government came in the form of further licences to expand.[9] Without such licences forthcoming from the government, a chaebol could not expand even though it might have a very strong desire to do so. In addition, if a chaebol chose to enter an industry that was particularly risky, it too was rewarded with licences, which did not just give it permission to expand, but more importantly, signified official approval to make vast profits. More often than not, these licences were permission to expand into sectors known to be highly lucrative.[10] Overall, the performance and incentive based system greatly encouraged the privately run chaebols to diversify and expand on an unprecedented scale, which accounts for the fact that the current leading chaebols, Samsung, LG, and Hyundai are now multi-billion dollar enterprises.

In cases where a chaebol has run into problems, often because of poor management, the government has shown itself willing to wield the stick by letting it go bankrupt. The advantage for the government is that it gives the system credibility and ensures that only successful firms are allowed to survive. In essence, this is a Korean version of the 'visible hand' of the state, albeit a very different version from the Japanese 'invisible hand' which Tyson (1992) has alleged exists. A chaebol facing bankruptcy, even though it might be in a sector that is in a healthy state, stands almost no chance at all if the government is determined to let it go under. When Park came to power, the government ensured that no large enterprise could turn to anyone else except the state for financial assistance because the whole banking system was nationalised.[11] Under external pressure from the US, in 1983 the government privatised the commercial banks. Although privatised, the government retained indirect 'control' because they did not become independent in the Western sense. Instead their ownership was transferred from the government straight into the hands of the chaebols.[12] This way the government could maintain its overall control of the financial system.

In the majority of cases where a chaebol has run into difficulties, perhaps due to problems outside of its control, the government has stepped in to bail it out. The government has got into itself into quite a serious trap. On the one hand, it is occasionally prepared to let a chaebol go bankrupt to give its system of performance credibility, but on the other, it cannot afford to let too many of the chaebols go under because of the repercussions it would have on the rest of the economy. Some of the chaebols have become so large that if they were allowed to go under the government would almost certainly be faced with considerable economic chaos which it would be unable to control. In 1984, for example, the three largest chaebols alone accounted for 36 per cent of the country's GNP.[13] Recognising the extent of the problem it had brought about, the government in the 1980s sought to constrain any future growth of the chaebols, but it was unsuccessful. Contrary to the wishes of the government, the chaebols actually grew in size, thereby exacerbating the problem of economic concentration in the hands of a few corporate giants. The government provided the chaebols with an additional incentive to grow when it transferred the ownership of the commercial banks to them. With easy access to considerable financial resources, it propelled the chaebols to go on a major spending spree. They proved to have a particularly voracious appetite for state enterprises that were being privatised and other firms in financial difficulties, which caught their attention.[14]

This leads us on to the question why have the chaebols become so successful in LCDs? Once again, one has to look at the role played by the government, which has been instrumental in enabling the chaebols to learn about technology, and at the strategies used by them in the semiconductor industry. A very well known fact about technology is that the more advanced it becomes, the more expensive it is to develop. In South Korea's case, when the government assumes the role of banker it does far more for the chaebols than simply bail out the occasional one that has run into difficulties. Due to the government's enormous influence over them, it set the agenda that it wanted them to follow and adhere to. The chaebols' first priority was investment rather than the pursuit of profits. However, to make it possible for the chaebols to have the capital to invest on the scale it wanted them to, the government needed to skew the financial system.[15]

For economic development to advance there has to be sufficient capital available for long-term investment. The normal source of this type of capital is domestic savings, but in South Korea these were insufficient to meet the country's requirements and the government chose to heavily subsidise selected chaebols and to borrow from abroad. Rather than coercing the chaebols to borrow extensively from the nationalised banks at commercial rates of interest, the government allocated long-term credit at negative real interest rates in order to stimulate specific industries.[16] In the 1970s the government controlled over two-thirds of the aggregate investment funds available to the economy[17] and with it the country experienced some of the highest rates of investment seen anywhere in the world. Between 1962–72 the ratio of investment to GDP averaged 21.1 per cent, and this increased to 29.6 per cent between 1973–79, before levelling off at 30 per cent between 1980–89.[18]

One of the main ways the chaebols have been able to fund their investment programmes is through easily obtained credit from the state banks and public funds.[19] At their disposal are a number of financial bodies created under the Technology Development Promotion Act of

1972. The main ones are the Technology Development Fund, Industrial Development Fund, Korea Development Bank, and the Technology Development Reserve Fund.[20] In return for this huge financial support the government not unnaturally expected the chaebols to fulfil their obligations. The mere prospect of this support being cut off was sufficient to ensure that those selected chaebols did their utmost to maintain their performance and compete with each other on grounds other than price.

The acquisition of technological capabilities

The foundations of the semiconductor industry in South Korea were laid back in the 1960s, as a result of investment by US manufacturers who were eager to take advantage of cheap labour to assemble discrete devices that were then later imported for use in the domestic market.[21] From the perspective of US firms they were doing nothing more than taking advantage of low labour costs offered by South Korea to help maintain their competitiveness, but from South Korea's perspective this was an early form of technology transfer. This kind of investment initially did little, if anything at all, to help South Korea develop its own indigenous industry. In quantitative terms, the investment from the US gave the industry an important boost in that at least got it off the ground. According to Kim (1996) in 1969 semiconductor exports amounted to $35 million, representing 5.6 per cent of total Korean exports.[22] Virtually all exports, however, were made up of the production of firms like Fairchild and Motorola, who accounted for approximately 95–99 per cent of Korean integrated circuit (IC) production.[23]

In the 1970s the US firms in South Korea were joined by Japanese firms who were just as eager as their US competitors had been a decade earlier to take advantage of the country's cheap labour. The semiconductor industry continued to grow, but it remained dependent upon investment by the foreign firms that dominated it. In addition to this, efforts to develop a domestic industry started to gather pace when a group of US-based South Korean expatriates set up a joint venture, South Korea Semiconductor, with Samsung Electronics Group (part of the Samsung chaebol) in 1974. This was the very first Korean wafer-processing plant, and was the catalyst for the development and production of CMOS Large Scale Integrated Chips (LSI) in Korea for the very first time.[24] Because of financial problems, ownership of the joint venture was passed to Samsung in 1978, as part of the newly established Samsung Semiconductor and Telecommunications Company (SSTC).[25] After Samsung's move into the semiconductor industry, it was duly followed by LG (formerly known as Lucky Goldstar) in 1979, when Goldstar Semiconductor was set up.[26] Apart from the influx of Japanese investment, and the tentative moves by Samsung and LG to set up their own semiconductor operations, the industry saw little change throughout the 1970s, but all that changed in the 1980s which saw the industry take off in its own right. In particular, it was the strategies pursued by Samsung and Hynix (formerly known as Hyundai) and favourable market conditions in the second half of the decade which led to a complete transformation of the industry.

Strangely enough, neither the government nor the chaebols had viewed the semiconductor industry as being a 'strategic' industry prior to the 1980s, one that would have been singled out for special treatment. Kim (1996) points out that it was not until the early 1980s that the government properly appreciated how important the semiconductor industry actually was. This was despite the enactment in 1969 of a law promoting the electronics industry and an eight-year plan for the industry.[27] The reason why the government overlooked it was because during the 1970s, it had focused its efforts on building up capacity and developing the chemical, steel, machinery and shipbuilding industries.[28] Once the government had realised the economy stood to gain enormously from an indigenous semiconductor industry, it swung into action. This is the key to explaining the initial lag and subsequent catch-up in the industry. The industry was included in the fifth Five Year Plan starting in 1982, as one of the principal 'target' industries.[29] This was promptly followed in 1983 by the formation of the Semiconductor Industry Fostering Plan which removed the import duty on production equipment for the industry.[30] As one might expect, the government assisted the chaebols to get themselves established in the industry. The government gave them enormous financial backing, but there were strings attached to this, and they had to export almost from the outset. To avoid falling foul of the conditions imposed on them, the chaebols had to put production at the centre of their strategies. However, for them to excel in this area they had to look overseas for assistance.

One of the first steps Samsung, LG and Hynix took to develop their technological capabilities was to enter into a wide range of technology transfer agreements, mainly with firms in the US.[31] Between 1983–85 Samsung concluded eight agreements, five of which were with US firms.[32] During 1981–85 Hynix concluded five agreements which were all with US firms, and LG concluded seven agreements, also all with US firms.[33] The great advantage of these agreements for Samsung and LG was that they enabled them to obtain a wide range of technology that was absolutely vital to developing their design, fabrication and production capabilities. The agreements covered a wide range of semiconductor technology including, DRAMs, SRAMs, NMOS & CMOS technology, gate arrays and EEPROMS. It was precisely the kind of technology that had not been forthcoming in the investments made by the US and Japanese firms in the 1960s and 1970s.[34] As well as obtaining vital technology necessary for building up expertise in production, the chaebols increased their investment on a huge scale.

Investment was not only channelled into R&D,[35] but also into expansion of production at home and abroad. All three chaebols established subsidiaries in the US, Samsung and Hynix in July 1983, followed by LG exactly a year later. There were significant benefits for the chaebols by expanding their presence abroad as it gave them immediate access to advanced technologies and major markets.[36] The other urgent necessity for the chaebols, access to major markets, was important for them in order to boost their exports, the most stringent of the government's requirements imposed on them in return for financial support.

When Samsung entered the semiconductor industry in 1983 it decided to focus on DRAMs for three principal reasons: the large size of the market; the chaebol had earlier seen other 'latecomer' firms (Japanese firms) catch up in the sector; and the DRAM's relatively simple design would not act as a barrier to competing with other firms so long as it could master the production technology.[37] Once it had committed itself, Samsung moved very quickly and announced in November 1983 it had successfully developed the 64K DRAM. Although Samsung had already started work on the 64K DRAM earlier in that year, prior to the decision being taken to focus the chaebol's resources on the technology, it was the speed with which Samsung had learned to master the technology which took the industry completely by surprise. And that was not all: in the following year, Samsung built a production facility, commenced mass production, and then proceeded to export to the US.[38]

This was to prove a major turning point to the extent that it showed Samsung had finally come of age in electronics. Not long after its success in the 64K DRAM, it made two more announcements that sent further shock waves through the industry, confirming beyond any reasonable doubt that Samsung had emerged as a major competitive threat to the other more established firms. The first announcement was to say it had successfully developed the 256K DRAM, confirming that its previous success in the 64K DRAM was not a 'one-off'. The second, was perhaps even more significant, it had developed the next generation of DRAMs, the 1M DRAM, on its own. This signified it was in a position to carry out its own product development and could now master the appropriate DRAM design technology.[39] Since the mid-1980s the chaebol has continued to forge ahead. By 1990 it had become the world's leading DRAM producer,[40] and has maintained that position since. By the third quarter of 2004, its global market share in DRAMs stood at 31.4 per cent.[41]

Samsung's rapid rise to the top of the industry's ladder has rested principally on it being able to acquire all the DRAM technology it needed from the US, and focusing on incremental product and process innovations. The chaebol's emphasis on the production processes was a tacit acceptance of its continuing dependence on foreign semiconductor equipment and materials. To make the most efficient and effective use of all the equipment and materials it was importing, Samsung had to develop a thorough knowledge of the whole production process, and continuously look for ways of improving it. An indication of just how dependent the chaebols like Samsung have been on foreign imports can be found in Hobday (1995). Throughout the 1980s and early 1990s, the country's semiconductor industry imported nearly 75 per cent of all the equipment and materials needed for production.[42] The strategy of importing, assimilating and integrating foreign technology allowed it to achieve higher yield levels than its Japanese competitors.[43] One other factor that significantly contributed to Samsung's meteoric rise during the 1980s was the huge change in the market conditions within the semiconductor industry.

In Chapter 4 it was noted that the effect of Japan catching up with the US in semiconductors led to the first Semiconductor Trade Agreement (SCTA), so that the industry could be better 'managed'. The US industry had been particularly hard hit by the severe downturn during 1985–86, which forced many US firms to permanently halt the production of DRAMs. One of the objectives of the SCTA was to try and prevent another repeat of this occurrence and give US firms the necessary leeway to recover. As the SCTA was only between the US and Japan, when the market started to improve in 1987, Samsung was left completely free to fill the vacuum created by the absence of the US firms.

To develop its technological capabilities, Hynix followed a strategy more or less identical to Samsung's. In 1983 it set up both a research institute in South Korea and a subsidiary in the US, with the objective of importing technology to help with its efforts to develop its technological base at home.[44] Where Hynix differed slightly from Samsung was in its choice of memory chips, it opted for the SRAM (static RAM), unlike Samsung which had chosen the DRAM. The distinction between the two types of chips is that the SRAM is far more technologically complex than the DRAM. Hynix chose this particular area, as it was one where there was little Japanese competition. As a newcomer, there was no possible way it could have competed successfully against well-established Japanese firms in a fiercely competitive market. Hynix's wish to avoid competing with Japanese firms turned out to be a mistake; it found producing SRAMs far more difficult than it had anticipated. It ran into problems because of faults with the complex design of the chip it had developed, and this prevented Hynix for about a year from being able to achieve satisfactory yield rates of 16K SRAMs. These difficulties with the SRAM led to a slight alteration in its strategy: Hynix decided to retain its commitment to the SRAM, but in December 1985 it decided to diversify into DRAM production as well.[45]

The timing of Hynix's decision to get involved in DRAMs put it at an immediate disadvantage to Samsung. To catch up with Samsung, Hynix – with no time to spare to develop its own chip designs – had to import them from the US, and became a subcontractor for two US firms, General Instrument and Texas Instruments. The strategy did not work out for Hynix in the way it had perhaps been expecting. Unlike Samsung, which had successfully mass-produced DRAMs using designs imported from abroad, Hynix's chip designs-obtained from the US firm, Vitelic, proved to be completely unsuitable for mass production. Except for the 64K DRAM, yield levels were far below the industry norm, for example, yield levels for the 256K DRAM were below 30 per cent, which led Hyundai to incur huge losses.[46] On the other hand, producing DRAMs, especially for Texas Instruments worked out well for Hynix. Between 1986–91 it produced the 256K DRAM for the US firm and in the process built up its development and production capabilities. The DRAM that Hyundai produced for Texas Instruments turned out to be very successful in two aspects. Firstly, it greatly helped improve the chaebol's very poor financial position, and secondly, it allowed Hynix to develop its own 256K DRAM. The way that Hynix was 'forced' to establish its presence in semiconductors, and the problems it encountered, cost it an awful lot of money which it could ill afford. Due to both the internal and external pressures it had been under to get the production side of its operations right, it made remarkable progress along the learning curve, and achieved the highest yield rate of any of the South Korean semiconductor producers.[47]

Throughout the 1990s Hynix continued to invest heavily in DRAMs and made rapid progress up the technological ladder. Due to the conditions of its agreement with Texas Instruments, Hynix's performance in the sector was constrained until around 1992.[48] It continued to produce 1M DRAMs for Texas Instruments on a sub-contractual basis, but went on to independently produce the 16M and 64M DRAM.[49] LG was the last of the leading chaebols into DRAMs as it was far more cautious than either Samsung or Hynix when it came to making large investments of a risky nature. Unlike Hynix, which was a newcomer at the time of its decision to enter the semiconductor industry, LG was already producing logic chips and 4-bit and 8-bit microprocessors when it made its decision to expand into DRAMs in the late 1980s. By having a relatively diverse product range, the strategy of LG would suggest that it was more comfortable with its investments spread over a number of areas, thus reducing its exposure to cyclical trends and Japanese competition. What finally seems to have enticed LG into DRAMs and copying its domestic rivals was not the prospect of supernormal profits, but the huge potential for expansion, a key priority of any chaebol. An additional plus was that DRAMs were a complementary investment which fitted nicely into LG's existing portfolio.[50]

One might have expected that LG's prior experience in producing other kinds of semiconductors would have been a major advantage when it decided to start producing DRAMs. Surprisingly though, LG found its technological capabilities were deficient in precisely the same areas as Samsung and Hynix had been a few years earlier, DRAM designs and production technology. Following in the footsteps of the other two chaebols, LG had to look abroad for what it required. For its DRAM designs, LG had to enter into a number of technology licensing agreements, in a similar fashion to Hynix. Given the lateness of its entry, it simply did not have any time to spare to embark on a programme to develop its own. For its production technology, LG did not turn to the US but to Japan's Hitachi, perhaps a reflection of the change in technological leadership that had occurred in the industry by then. Like Hynix, it took LG some time to make an impact on the industry.

Its difficulties were caused in part by its slow progress in building up its own development capabilities, and in part by the increasing reluctance of foreign firms to continue to license their technology. In view of the success of Samsung and Hynix, Hitachi and Motorola both refused to extend the agreements that they had with LG to produce 16M DRAMs.[51] Hitachi and Motorola had assisted LG to gain a foothold in DRAMs by supplying it with the technology for 1M and 4M DRAMs, but this was as far as they were prepared to go.[52] By refusing any further cooperation with LG, Hitachi and Motorola used a blocking strategy, which they hoped would radically slow down the chaebol's progress by dramatically increasing the costs of the technology it would be forced into purchasing. By 1999 LG had become the fifth largest DRAM producer in the world with an 8.4 per cent share of the global market, but its rapid progress was brought to an abrupt end when chronic financial pressures forced it to sell its DRAM operations to Hynix in the same year.[53]

The 'latecomer' firm

Earlier on, the concept of the 'latecomer' firm was mentioned very briefly without any explanation as to what it actually was. In the current context, the success of the chaebols in semiconductors would imply that a latecomer firm is one that decides to enter a specific industry at a particular time, and then proceeds to catch up with the leading firms. For a latecomer firm to be successful, though not necessarily on the scale of Samsung, it has to overcome major handicaps and has few distinct advantages. The main feature of the latecomer firm, according to Hobday (1995), is that it is typically located in a newly industrialising country (NIE). By the very nature of its location, the latecomer firm suffers from two major handicaps. The first arises from its detachment from the main centres of science and innovation, so it is automatically constrained by a poorly developed national technological infrastructure which does not possess the research, development and engineering capabilities of the advanced industrialised countries. The second concerns the international markets in which it wants to operate. More often than not the latecomer firm suffers from considerable isolation from them.[54]

From the standpoint of the chaebols, these handicaps were all very real. The geographical location made overcoming existing technological barriers extremely tough. To have any chance of solving the problem, the chaebols had to use many of the channels open to them to acquire the technology that they did not possess at home. As far as establishing a presence in international markets was concerned, it was far from easy. To put themselves in a position to compete effectively the chaebols had rapidly to learn to build up their production capabilities. As a typical latecomer firm, Hynix's early experience with SRAM technology shows that unless the production process is properly understood a firm's strategy stands a very good chance of falling apart very quickly.

On its side, the latecomer firm has two important advantages, which if used effectively can help gain it a useful early competitive advantage. The first relates to the stock of technology it has at its disposal. Samsung exploited this advantage to its absolute maximum when it was looking for an 'easy' way into semiconductors. It had the luxury of being able to choose from a variety of DRAM designs, developed primarily in the US, and select the one(s) that it was confident of being able to put into production. The second advantage the latecomer firm has is financial. As a chaebol is centrally organised and structured in a very hierarchical fashion, it is in a very good position to use cross-subsidisation to finance the expansion in a sector in which it wants to build up a significant presence.[55] This is made possible because the number of different affiliated firms that make up a chaebol are grouped around the chairman and central office, which have direct control over how a chaebol's resources are allocated.[56] The centralised structure of a chaebol enables decisions to be taken quickly.[57] Once a decision has been made it can then be worked out among the various firms of the chaebols which will be involved in supporting the proposed expansion.[58]

A glimpse of how important this can be was vividly displayed in the severe downturn of the mid-1980s which saw the chaebols fill the vacuum caused by the exodus of US firms out of DRAMs. As the average price for a 256K DRAM hovered around the $2–3 level between the third quarter of 1985 and the third quarter of 1987,[59] Samsung was able to continue investing in product development and production capacity even when it was in the red to the tune of $250 million at the end of 1986.[60] This was made possible because Samsung was prepared to transfer substantial amounts of capital from Samsung Electronics and its telecommunication division, which was then profitable.[61] The two other chaebols did likewise. The effect of this was that it helped South Korea capture approximately 15 per cent of the global DRAM market by 1989[62] at the expense of Japan, which saw its share of the global market decline from approximately 80 per cent in 1986 to 57 per cent in 1991.[63]

LCDs

With the correct government backing, sufficient financial resources, and access to the appropriate technology, the chaebols as latecomer firms have demonstrated beyond question that it is possible to move late into an industry that is continuously driven forward by technological change, and succeed.[64] What the chaebols have achieved in DRAMs they have repeated in LCDs. They have achieved this by employing a very similar strategy to the one they used in the DRAM sector, but with one major distinction. In LCDs, they started at the bottom of the technological ladder, whereas they did not have to in DRAMs. In DRAMs, the chaebols were very fortunate to be in the position where they could buy chip designs, particularly from the US, that were easy to put into production. In LCDs, this has not been an option that has really been open to them. This has been all but closed off to the chaebols because Japan is the dominant player in the industry, and it does not have the tendency of selling its technology to its competitors on a scale anywhere like the US does.

Basic research on liquid crystals has been carried out in universities and government-funded institutions in South Korea since the late 1970s and on amorphous silicon TFTs since 1980. [65] As with the semiconductor industry, there were the roots for an indigenous industry, but they have taken some time to grow and become strong enough for the industry to become sufficiently competitive at a global level. The origins of the industry can be traced back to 1979, when two firms, Seotong and Handok, started to produce TN-LCDs for watches and calculators,[66] in the same way as the US and Japanese industries got off the ground. This was around the time when the turmoil in the digital watch industry had dealt the fatal blow to the US LCD industry, but had created a wonderful window of opportunity for the Japanese LCD industry. With the collapse of the US industry, the only firms that the chaebols would eventually have to compete with were Japanese ones. In time honoured tradition, the chaebols followed the same strategy as the Japanese firms used, which meant that they should be able to compete effectively with them, assuming nothing went seriously wrong anywhere along the line.

From the early 1980s, Samsung and LG produced TN and then later STN-LCDs, but it was not until after 1985, that they began to appreciate the significance of what was gradually becoming a TFT bandwagon. In their analysis, Farrell & Saloner (1985) show with the use of a model that if all users (firms) have perfect information about a specific standard (or technology) that would make them better off, they will all adopt it.[67] The attractive features of the standard entice more and more users to adopt it, which then starts a bandwagon rolling. Alternatively, in a case where there is incomplete information, there is likely to be some form of inertia inhibiting adoption of the standard. Farrell and Saloner identify two forms of inertia, 'symmetric' and 'asymmetric'. With the former, the users are unanimous in their preference for the new standard and yet they do not make the change,[68] because they are reluctant to start the bandwagon rolling for one reason or another. Asymmetric inertia, on the other hand, occurs when users differ in their preferences over technologies, and the band-wagon for a particular standard can be very slow to get rolling because those users in favour of it are not sufficiently numerous.[69]

Because of the widespread recognition within the industry of the need for a completely different addressing system to replace the one used in STN-LCDs, the number of Japanese firms committing them-selves to developing TFT technology was growing. They knew that without a new addressing system, the LCD could not produce crystal clear images of complex graphics and moving pictures, thereby severely limiting its future applications. Among Japanese firms there was asymmetric inertia, as there were technologies available other than TFTs that could potentially have been used for the new addressing system. Two of these LCD technologies were the ferroelectric LCD and diode LCD. Although these technologies were competing with the TFT, it was the TFT that was the most likely of the technologies to be adopted, even though the eventual users did not have 'perfect informa-tion' about it.[70] Bahadur (1983) points out that at the Society for Information Display (SID) Conference in 1982, there were a lot of dis-cussions on the TFT approach.[71] By then a great deal was already known about how the TFT system worked, the only factor which cast a shadow over its prospects was the problem of yields. No one was as yet sure how a firm could produce a TFT-LCD on a commercially viable basis given the excessive complexity of the production process. Because overall the TFT system was superior to its rivals, Bahadur notes that once the yield problem had been overcome, it was destined to become the dominant form of LCD.[72]

By the late 1980s, when notebooks were coming on to the market in ever-increasing numbers, the TFT bandwagon had really begun to roll. As the Japanese firms' understanding of TFT technology improved, and the displays proved themselves on the mass market, the chaebols knew they too would have to join the bandwagon if they were to take full advantage of the opportunities with which the TFT-LCD presented them. Unlike in semiconductors, where Samsung took the lead, this time round it was LG. It jumped on the bandwagon in 1987 when it commenced research on TFT-LCDs and was followed by Samsung in 1991. Whereas LG had been the most cautious of the chaebols in DRAMs, it is Hynix that has been the most cautious out of the three in TFT-LCDs. It did not finally join the bandwagon until 1993, but was forced to exit the industry in 2001 owing to severe financial pressures.[73]

Now that two of the chaebols have been fully committed to TFT-LCDs for more than a decade, it is possible to assess their individual progress to date. Despite LG starting research on TFT-LCDs four years earlier than Samsung, it is Samsung that has really forged ahead. The success it has enjoyed in DRAMs, it is now enjoying in TFT-LCDs. Within a period of approximately seven years it has become one of the world's leading producers. In 1997 Samsung was ranked between fourth[74] and sixth[75] in the world. In March 1999, it was reported that Samsung had become the world's leading producer in terms of market share with 17.5 per cent, surpassing Sharp, which has been the leading firm in the industry since the early 1970s.[76] With an initial investment of $375 million, it achieved this by a highly adept move from research to the production stage.[77] In 1991 it set up a pilot production line and an R&D team that has grown steadily from 20 to 100. By 1993 Samsung had developed a 9.5-inch TFT-LCD, and in the September of that year it increased the pilot line's monthly output of TFT-LCDs. Its rapid progress made it viable to start the construction of its first commercial production line in December 1993. This was completed in October 1994 and mass production started in 1995.

The speed at which Samsung moved into mass production was facilitated by its strategic alliance with Fujitsu, formed in April 1995. Unlike in DRAMs where Samsung imported chip designs from the US which it could put almost immediately into production, in TFT-LCDs it has found that it has not been able to buy some of the technology it needs straight 'off-the-shelf'. Samsung and Fujitsu agreed to share important TFT-LCD technologies.[78] Simultaneously, Samsung made rapid strides in the development of larger size TFT-LCDs. In February 1994, four months after the production line had been completed, Samsung developed a prototype 12.1-inch TFT-LCD, which was followed by a 14.1-inch TFT-LCD, another four months later.[79] Samsung's advance has continued unabated developing larger and larger displays from 13.3-inches and beyond, capturing 23.3 per cent of the global market in the process.[80]

Considering the extensive time gap between when research started in 1987 and the beginning of mass production in 1996, it appears that LG has found that competing in TFT-LCDs is much harder than in DRAMs, even with its first tranche of a long-term investment programme of $500 million. [81] In 1990 it set up a laboratory specifically for TFT-LCDs, and it took the chaebol another two years to successfully develop a wide range of prototypes.[82] These included not only the usual 9.5-inch, 10.4-inch, and 12.1-inch TFT-LCDs, but also much smaller ones as well, for example, 3-inch, 4-inch and 5.6-inch.[83] It may have taken LG a while to develop a number of different sized LCDs, but it confirms that the chaebol has a proven capability in research and is now nowhere near as reliant as it once was on imported technology.

In contrast to Samsung, which built a pilot production line at the same time as it was developing its prototypes, LG only commenced production using a pilot line in 1994, after it had developed its prototypes. LG finally commenced mass production in June 1995 when its production facility was completed. The facility had a capacity to produce 500,000 10.4-inch TFT-LCDs, the most important sector of the market in 1995, as this was where the demand for TFT-LCDs was then at its strongest.[84] As with Samsung, LG has had to look overseas to acquire some of the technology it needed to facilitate its entry into TFT-LCDs. In May 1991, LG brought an 8.7 per cent stake, worth about $1 million, in a US firm, Photon Dynamics. Through this purchase LG gained access to important TFT-LCD testing equipment and other technologies related to the display. In addition, LG formed a strategic alliance with the Japanese firm, Alps Electric, in July 1994. The alliance is a 50:50 partnership between the two firms who agreed to set up a research facility to develop the next generation of TFT-LCDs.[85] Even though it has not matched Samsung's very impressive performance, it has still managed to become the world's second leading firm with a global market share of 19.9 per cent in 2003.[86]

On its way to becoming one of the world's leading firms it should be pointed out that LG has experienced some major financial problems. The turning point came in 1997 when the industry was hit by a collapse in prices caused by a glut. In the first quarter of 1997, LG forecast that its revenues from TFT-LCDs in 1998 would surpass $930 million, and these would then more than double in 1999 to $2.3 billion.[87] In the event LG's forecasts proved wildly over-optimistic.

In 1998 LG's actual revenues from TFT-LCDs were $500 million, little more than half what LG had expected a year earlier.[88] Although prices and demand started to recover in the third quarter of 1998, 1999 was not the 'bonanza' year that LG had originally hoped it would be. LG believed that its revenues from TFT-LCDs would reach $1.8 billion, a three-fold increase on the previous year.[89] In August 1999 it was reported that up to the end of the second quarter of that year LG's revenues had only reached $751 million.[90] The loss of revenue that LG had expected to generate in 1998 and 1999 took a heavy toll on the chaebol. LG was forced to seek a foreign partner to maintain its position in the TFT-LCD industry because it did not have the capital available to continue to invest in TFT-LCDs at the rate it thought appropriate. After searching around, LG concluded an agreement with Philips, which paid $1.6 billion for a 50 per cent stake in LG's TFT-LCDs operations.[91] This provided LG with the additional $1.2 billion it needed to expand its TFT-LCD production. Its partnership with Philips has proved a major success.

So far we have only analysed the tactics of the individual chaebols, but there remains a very important question, which is inextricably linked to their success: what has been the role of the government? To all intents and purposes, the government has adopted the same attitude as it took when the chaebols made their move into semiconductors. Whilst it has openly encouraged the chaebols to take what have been huge risks, it has maintained its traditional close relationship with them. Behind the scenes, the government has supported the chaebols by coordinating research programmes run by three government departments, the Ministry of Trade, Industry, and Energy (MOTIE), the Ministry of Information and Communication (MOIC), and the Ministry of Science and Technology (MOST). The government has supported around 40 display development projects, emphasising manufacturing technology development in information display areas.[92] The LCD projects that it supports fall into three main categories: modules, materials and components, and equipment. This has been backed up by financial support in the form of loans and contributions that have been distributed amongst fourteen different firms,[93] which presumably includes the leading chaebols. Between 1993–97 support from MOTIE, MOIC and MOST for the development of LCDs and other flat panel display technologies amounted to $63 million.[94] The strategy of the government has been to continue to provide substantial long-term support to the chaebols. It has used measures such as the G-7 Highly Advanced National (HAN) Projects, in which MOTIE was scheduled to invest $280 million between 1996–2000.[95] The participating firms in the project(s) would gain from financial support in the form of direct financing and tax exemptions. One of the main objectives of the research programme was to assist in the development of large displays – a 27 inch TFT-LCD, which has now been achieved.[96]

One final measure to which the government has given its full support has been the formation of the Electronic Display Industrial Research Association of Korea (EDIRAK) in May 1990. The original idea came from the leading chaebols who were convinced that a coordinated approach to the development of LCDs and other flat panel technologies, along the lines pursued by Japanese firms for so many years, could only be beneficial. The function of EDIRAK is to jointly develop and standardise information display technologies through the distribution of government funds to firms for industrial R&D.[97] The government through MOTIE provides 40 per cent of the budget and the firms contribute the other 60 per cent. The other two government departments also have a role in the functioning of EDIRAK, mainly in supporting aspects related to basic research and education.[98] Exactly how effective EDIRAK has been is almost impossible to quantify, but the government at least believes it has and will continue to make a valuable contribution in the future, as MOTIE is expected to remain as one of its major sources of funding.[99]

Taiwan

With the failure of the US and Western Europe to catch up in TFT-LCDs it looked as though the industry would remain divided between Japan and South Korea. However, Taiwan rapidly emerged as the 'dark horse' of the industry. From early on, it was clear that Taiwan possessed the technological capabilities to become a major player in TFT-LCDs, and it has now moved into a position where it can show its true colours.[100] What enabled Taiwan to realise its potential and become a formidable competitor, amongst other things, is its vibrant DRAM industry, its strong presence in consumer electronics, and the country's willingness to take full advantage of what Rosenberg (1982) has observed is inevitable, the transfer of industrial technology.[101] The context in which Rosenberg firmly believes the transfer of industrial technology is inevitable is from the developed countries to the less developed countries. Increasingly, it is looking as though Rosenberg's observation is in need of updating. In his analysis he seems to have overlooked the possibility of significant transfers of industrial technology between developed countries. In TFT-LCDs, this is precisely what has happened. Taiwan initially imported from Japan older generations of TFT-LCD technology rather than the latest generation technology available.[102] Now that has started to change. Certain Taiwanese firms have joined forces with their Japanese counterparts to development advanced TFT-LCD technology.[103]

Linden *et al.* (1997) point out that although Taiwan's investment in TFT-LCDs lagged several years behind South Korea, this did not mean that Taiwan's late entry strategy would in any way preclude it from becoming a major player.[104] Both Taiwanese firms and the government wanted to enter the TFT-LCD industry. What Taiwanese firms wanted above all else was to substantially reduce their dependence on TFT-LCDs imported from abroad; at the same time they wanted access to reliable supplies.[105] As major producers of notebooks and other similar portable products, it is understandable why Taiwanese firms wanted to do this. For example, in 1995 they imported an estimated 1.9 million LCD units of various kinds for assembly in notebooks.[106] The government was equally keen that the country's firms should make a successful entry into TFT-LCDs because of the substantial negative impact it threatened to have on the country's balance of payments with Japan. In 1996 Taiwan's trade deficit with Japan was just under $14 billion, and TFT-LCDs accounted for approximately 10 per cent of this deficit.[107]

Taiwanese firms might have had the best of intentions in entering the TFT-LCD industry, but the speed of their entry has been constrained by a number of obstacles, such as short-term financial considerations and difficulties in raising capital for volume production.[108] These are very similar to the types of problems which Peter Brody experienced when he tried to interest US venture capitalists and US firms to develop TFT technology. The seriousness of the problems for Taiwanese firms was such that by the fourth quarter of 1997, no Taiwanese firm had started high volume production of notebook sized TFT-LCDs,[109] although Unipac Optoelectronics had been mass-producing 4-inch TFT-LCDs since 1994.[110]

Taiwanese firms have not let these obstacles stand in their way in TFT-LCDs. Taiwanese firms have proved themselves to be highly adept at overcoming the handicaps they faced at home. The solutions to many of their problems have been found through their decision to pursue strategies which have involved forging alliances with Japanese firms to gain access to TFT-LCD technology and production equipment, so that they could gear themselves up for mass production. For example, in April 1997 Chungwha Picture Tubes (CPT) signed a deal with Advanced Display Inc (a joint venture in Japan between Asahi Glass and Mitsubishi) to acquire TFT-LCD technology.[111] And in May 1999, Quanta Computer Inc. formed a joint venture with Sharp, called Quanhuei Optoelectronics Inc. Quanta owns 33 per cent of the joint venture and Sharp 10 per cent.[112] This is a trend that currently shows no sign of abating.

The transfer of Japanese technology has given the Taiwanese TFT-LCD industry the chance to take off after a very slow and difficult start. Taiwan has risen to the challenge to develop its own production and technological capabilities in TFT-LCDs and put the industry on a sustainable footing. The two principal reasons why Japanese firms have been eager to seek partners overseas has been to help stem the advance of the chaebols, and the lack of sufficient capital for investment. According to Fuller *et al.* (2003) since 1997 at least six Taiwanese firms have built fabrication facilities, and others have the technology to do so.[113] With the influx of firms into TFT-LCDs and investment totalling more than $9.4 billion according to Hung (2002),[114] Taiwan's market share has increased rapidly, from less than 10 per cent in 1999[115] to over 20 per cent in 2003[116] to an impressive 36 per cent in 2005.

Compared to the South Korean government, the Taiwanese government has adopted a much more hands on approach to the promotion and development of the country's TFT-LCD industry. Where the Taiwanese government's approach differs from the South Korean government's has been in its active attempts to bring the country's firms together in the form of an alliance to manufacture TFT-LCDs. Furthermore, the government has tried to persuade the firms to license TFT-LCD technology from it. Despite having the best of intentions, the Taiwanese government's approach has proved nothing short of a disaster.

The main body that has been entrusted by the Taiwanese government with the task of promoting technology is the Industrial Technology Research Institute (ITRI). It has a microelectronics research laboratory, the Electronics Research and Service Organisation (ERSO), which has been successful in developing pilot production facilities that have later grown into full-scale production facilities.[117]

Government efforts to develop LCD technology started in 1986. Two years later, these efforts received a major boost, when the government launched its Optical Information Project (OIP), a research programme to develop a number of different technologies, which included LCDs.[118] In total, five areas of LCD technology were selected for development; microfabrication for thin-film transistors, integrated circuits (ICs) for driver chips, panel assembly, module packaging and related materials.[119] By 1989 ERSO had managed to develop the first 3-inch prototypes for LCD projection television, and by the early 1990s it had successfully developed notebook sized TFT-LCDs. In 1992 the government stepped up its efforts to accelerate the development of the country's TFT-LCD industry. Through its Industrial Development Bureau (IDB), which is an offshoot of the Ministry of Economic Affairs (MEA), the government appealed to all those firms developing LCD technology to pool their resources.[120] Up to this point everything seemed to be going according to plan for the government.

Little did the government know that from 1993 onwards, all its plans for an indigenous TFT-LCD industry would crumble before its very eyes. The first thing to go wrong for the government in 1993 was when ERSO proposed a public investment of $400 million in TFT-LCDs.[121] Linden *et al.* (1997) note that this proposal was rapidly shot down by the firms which ERSO now consults on a regular basis as part of a formal review process.[122] The second 'hiccup' to occur in that year, was when ITRI tried to bring those firms that were interested in developing an indigenous TFT-LCD industry together in the form of a private sector alliance. ITRI set up a meeting with the firms where it unveiled to them how it was going to proceed. In addition to trying to get the firms to forge a successful private sector alliance, ITRI planned to carry on funding important research at ERSO before letting the private sector take over.[123]

ERSO's plans gained the valuable support of a venture capital firm, Champion Consulting, which acted as the catalyst to bring together five firms, UMC, CPT, Nan Ya, Picvue and TSMC, to form an alliance known as the Mandarin Display Manufacturing Corporation (MDMC). As a sign of their determination to enter the TFT-LCD industry and commitment to MDMC, each of the participating firms paid $40,000 to MDMC.[124]

For MDMC to stand any chance of success, it had to overcome two major hurdles. Firstly, it needed its own TFT-LCD production facility. In spite of the very high cost, it does not appear to have caused MDMC too many headaches. With MDMC consisting of at least two reasonably large firms, CPT and Unipac, raising the necessary $500 million would not have been too difficult. Secondly, MDMC had to decide whose technology it was going to adopt. In theory MDMC had the choice of two technologies, but in reality it only had one. As Japanese technology was unavailable, MDMC could have adopted Philips double-diode thin-film technology, or alternatively, it could have chosen to wait until ERSO's technology was ready for commercialisation.[125] In a dynamic industry like TFT-LCDs, waiting for ERSO to get its act together was never really on the cards, so the only option open to MDMC was to license Philips technology. Not surprisingly, Philips did everything it could to try and persuade MDMC to license its technology, but in the end all its efforts were in vain. Part of Philips' problem was that it was trying to sell a technology that no other major firm was remotely interested in, and MDMC must have known that. Philips' failure to win over MDMC was not only a body blow to Philips itself, it inadvertently helped sow confusion within MDMC. Linden *et al.* (1998) note that splits developed within MDMC over which technology to choose. The seriousness of these splits were so bad that no technology was chosen in the end, causing MDMC to collapse.[126]

In 1995 ERSO made a second attempt to form a private sector alliance to manufacture TFT-LCDs, this time using its own 10.4-inch TFT-LCD technology.[127] Despite the failure of its first attempt, ERSO managed to persuade three firms from its earlier attempt that they should make one more final go of it. This second alliance comprised of CPT, Nan Ya, Acer and China Steel, who are reported to have each paid $150,000 to participate in the alliance and for the commitment of their engineers to the project.[128] ERSO's idea was that the alliance would continue to develop the technology it had developed, with PVI responsible for the development of the prototype displays. ERSO envisaged that once the alliance had mastered the production of TFT-LCDs at the pilot stage, it would then spin off the whole operation into a private joint venture. ERSO was hoping to repeat with TFT-LCDs what it has done on a number of occasions with integrated circuits: develop a pilot production facility that then evolves into a full-scale production facility.[129]

To its absolute dismay ERSO had its plans thwarted at the very last minute, this time by an uncooperative firm. PVI's management decided that the costs of participation in ERSO's venture would outweigh the benefits, and consequently refused to join the alliance.[130] With no means of compelling PVI to go along with its plans, and with the other firms in the alliance unwilling (or unable) to take PVI's place, this second alliance went the same way as the first and duly collapsed.[131]

Summary

The strategies adopted by South Korea and Taiwan to enter the TFT-LCD industry have proved highly successful. Their contrasting strategies could not have been more different, yet the outcome has been the same. The most unusual aspect of the strategies of South Korea and Taiwan is the level of government intervention. In a country famed for extensive government intervention in industry, intervention by the South Korean government on this occasion has been kept to a minimum. The responsibility for the development of the industry has largely been left in the hands of the chaebols. At the other end of the spectrum, Taiwan, which is not usually associated with extensive government intervention in industry, has adopted a more 'hands on' approach. As discussed earlier, to the government's surprise its approach did not work out as it had envisaged. Originally, the government assumed that if it could encourage Taiwanese firms to form an alliance, it could then direct the development of the country's fledgling TFT-LCD industry. What the government did not anticipate were the difficulties it would encounter along the way. As these difficulties proved insurmountable, it was forced to change tack. Rather than trying to direct the development of the TFT-LCD industry in the way that it sees fit, it now works in close tandem with the firms, consulting with them on a regular basis, and at the same time has allowed them to form alliances with overseas firms of their choosing. Although South Korea and Taiwan have very different NSIs, they have both had the technological capabilities to support the development of LCDs. Moreover, the firms have demonstrated their capacity to learn about the technology in a very convincing fashion.

Notes

1. Park, WS (1994) 'The Recent Status of LCD Development in Korea', *The Electromechanical Society Proceedings*, Vols. 94–95, p. 11.
2. *Ibid*:16. 'Korea's electronics giants face off in TFT-LCD market', *Electronics International*, 23 August 1993, p. 1.
3. Special Feature; 'Korea's TFT-LCD Business: Getting Brighter', *SEC Family News*, Samsung Electronics, September 1997, p. 8.
4. 'Korean Firms Leading TFT-LCD Market', Info Tech section, *Korea Times*, 18 February 1999 (www.korealink.co.kr).
5. Samsung Electronics, Special Report, *The Economist*, January 15 2005, p. 67.
6. For a concise history of the country's backwardness and the political turmoil it has been through, see Amsden, A (1989) *Asia's Next Giant: South Korea and Late Industrialisation*, Oxford: Oxford University Press, Chapter 2.

7. *Ibid*:16.
8. *Ibid*.
9. *Ibid*:14.
10. *Ibid*.
11. *Ibid*:16.
12 Kim L (1993), 'National System of Industrial Innovation: Dynamics of Capability Building in Korea' in RR Nelson (ed.) *National Innovation Systems: A Comparative Study*, New York: Oxford University Press, p. 368.
13. *Ibid*:116.
14. Kim, 1993:368.
15. For a detailed analysis of how extensive government intervention in the financial system has been, see Nembhard, JG (1996) *Capital Control, Financial Regulation, and Industrial Policy in South Korea and Brazil*, Westport, Connecticut: Praeger, Chapter 4.
16. Amsden, 1989:144.
17. Nembhard, 1996:72.
18. *Ibid*:75.
19. Kim, 1993:372.
20. Cheng-Fen Chen & Sewell, G (1996) 'Strategies in technological development in South Korea and Taiwan: the case of semiconductors', *Research Policy*, Vol. 25, No. 5, p. 761, footnote 6.
21. *Ibid*:763.
22. Kim, SR (1996) 'The Evolution of Governance and the Growth Dynamics of the Korean Semiconductor Industry', Working Paper No. 20, Sussex European Institute. Figures given in section 4.1.1, 'Foreign investment, hierarchical governance and export promotion policy in the 1960s'.
23. *Ibid*.
24. *Ibid*.
25. Chen & Sewell, 1996:764.
26. *Ibid*.
27. Kim, 1996. Section 4.1.1.
28. *Ibid*. This was a government initiative called the 'Heavy and Chemical Industry Promotion Plan', launched in 1973.
29. Chen & Sewell, 1996:763–64.
30. *Ibid*:764.
31. Details of the technology and which overseas firms the chaebols had agreements with are given in *Ibid*:764, Table 3.
32. Two of these agreements were with Samsung's US subsidiary, Samsung Semiconductor Inc.
33. One of these agreements was with AT&T, and also included a joint venture.
34. Chen & Sewell, 1996:764.
35. Between 1980–89 the chaebols' share of R&D investment increased to 83 per cent, while the government's declined to 17 per cent.
36. *Ibid*.
37. Kim, 1996: Section 4.2.1 'Entry and waiting for a 'window of opportunity': the efficacy of chaebol governance', sub-heading, 'Samsung's entry and strategy focused DRAM production'.

38. *Ibid.*
39. *Ibid.*
40. Hobday, M (1995) *Innovation in East Asia: The Challenge to Japan*, Aldershot: Edward Elgar, p. 62.
41. Samsung Electronics, Special Report, *The Economist*, 15 January 2005.
42. Hobday, 1995:63. Hobday notes that the country's very heavy reliance on foreign imports led the country to run persistently large trade deficits throughout the 1980s and early 1990s, especially with Japan. In 1991 alone imports related to electronics amounted to $11.2 billion (*Ibid*).
43. Kim, 1996: Section 4.2.1.
44. Kim, 1996: Section 4.2.1.
45. This section has been summarised from the section above in Kim (1996).
46. *Ibid.*
47. *Ibid.*
48. Chen & Sewell, 1996:768.
49. *Ibid.*
50. Kim (1996) Section 4.2.1.
51. Chen & Sewell, 1996:768.
52. LG had second-sourcing agreements with both Motorola and Hitachi (which included providing a turnkey facility for the production of 1M DRAMS).
53. 'Hyundai Emerges as World's Largest DRAM Maker', Koreabiz section, *Korea Times*, 12 March 1999. (www.korealink.co.kr).
54. Hobday, 1995:34.
55. Kim, 1996: Section 4.2.1.
56. *Ibid.*
57. *Ibid.*
58. *Ibid.*
59. Tyson, 1992:115, Table 4.4.
60. Kim (1996): Section 4.2.1.
61. *Ibid.*
62. Tyson, 1992:106. Figure 4.2. In the graph South Korea and Taiwan have been grouped together.
63. *Ibid*:126.
64. For a more extensive discussion on the concept of the latecomer firm, see Hobday, 1995, Chapter 3.
65. Kahaner, D (1994) Report on the Electronic Display Forum, Yokohama, Japan, April 1994 (www.atip.org/ATIP/public/atip.reports.94/flat.94.html).
66. Park, 1994:12. These firms later merged into Korea Electronic and Orion Electric, respectively.
67. Farrell, J & Saloner, G (1985) 'Standardisation, compatibility, and innovation', *RAND Journal of Economics*, Vol. 16, No. 1, Spring, pp. 70–83.
68. *Ibid*:72.
69. *Ibid.*
70. Arthur, WB (1989) 'Competing Technologies, Increasing Returns and Lock-In by Historical Events', *Economic Journal*, Vol. 99, pp. 116–131.
71. Bahadur, B (1983) 'A Brief Overview of History, Present Status, Developments and Market Overview of Liquid Crystal Displays', *Molecular Crystal & Liquid Crystals* Vol. 99, p. 348.
72. *Ibid.*

73. Pecht, M, Bernstein, JB, Searls, D & Peckerar, M (1997) *The Korean Electronics Industry*, New York: CRC Press, pp. 48–50.
74. 'Korea TFT-LCD Output will Rise Sharply', *Korea Economic Weekly*, 11 August 1997 (www.ked.co.kr).
75. 'Korean Firms Leading TFT-LCD Market', Info Tech section, *Korea Times*, 18 February 1999 (www.korealink.co.kr).
76. *Ibid.*
77. Pecht *et al.*, 1997:49. 'Korea's electronics giants face off in TFT-LCD market', *Electronics International*, 23 August 1993, p. 1.
78. 'For Money, Technology, Korean TFT-LCD Makers Knit Alliances', Korean Report, *Journal of Electronic Industries*, June 1996, p. 50.
79. Pecht *et al.*, 1997:48.
80. Samsung Electronics, Special Report, *The Economist*, 15 January 2005, p. 67.
81. 'Korea's electronics giants face off in TFT-LCD market', *Electronics International*, 23 August 1993, p. 1.
82. Pecht *et al.*, 1997:48.
83. Park, WS (1994) 'The Recent Status of LCD Development in Korea', *The Electromechanical Society Proceedings*, Vols. 34–35, p. 13.
84. *Ibid.* Pecht *et al.*, 1997:49.
85. 'For Money, Technology, Korean TFT-LCD Makers Knit Alliances', Korean Report, *Journal of Electronic Industries*, June 1996, p. 50.
86. 'Samsung Electronics', Special Report, *The Economist*, 15 January 2005, p. 67.
87. 'LG Enhances Leadership in TFT-LCD Production with $1 Billion Facility', LG TFT-LCD Business, March 1997 (www.lgsemicon.co.kr).
88. 'Philips targets screens for dazzling future', *Financial Times*, 19 May 1999.
89. *Ibid.*
90. 'LG Jacks Up Production Capacity to Meet Soaring Demand for TFT-LCD', Business section, *Korea Times*, 4 August 1999 (www.korealink.co.kr/times.htm).
91. 'Philips targets screens for dazzling future', *Financial Times*, 19 May 1999.
92. Pecht *et al.*, 1997:46.
93. *Ibid.*
94. *Ibid.*
95. *Ibid.*
96. *Ibid.*
97. *Ibid*:47.
98. *Ibid.*
99. *Ibid.*
100. Linden *et al.*, 1997:14.
101. Rosenberg, N (1982) *Inside the Black Box: Technology and Economics*, Cambridge: Cambridge University Press, p. 270.
102. Linden *et al.*, 1998:31.
103. 'AU Optronics teams with Fujitsu on LCD development', *Electronic Engineering Times*, 28 January 2003 (www.eetimes.com).
104. Linden *et al.*, 1997:14–27.
105. Pecht, M & Lee, CS (1997) 'Flat Panel Displays: What's Going On In East Asia Outside of Japan', p. 12. The old website from which this publication was obtained is www.calce.umd.edu/general/AsianElectronics/

Articles/Display.htm. The new website where details of this paper can be found is: www.calce.umd.edu/general/published/papers/article.htm
106. Pecht & Lee, 1997:12.
107. Linden *et al.*, 1998:25.
108. Linden *et al.*, 1997:27.
109. *Ibid*:14.
110. *Ibid*:15.
111. Pecht & Lee, 1997:14.
112. 'Quanta, Sharp in tech-license/joint venture deal', *Electronic Buyers' News*, 31 May 1999, Issue 1162 (www.techweb.com). The other firms to have formed alliance with Japanese firms are Chi-Mei which has formed an alliance with Fujitsu; Walsin Lihwa Corp and Winbond Electronics Corp have formed a joint venture known as Hannstar Display Corp with Toshiba; and Unipac has formed an alliance to co-develop a line of TFT-LCDs with Matsushita. 'Quanta close to forging FPD Deal', *Electronic Buyers' Guide*, 10 May 1999, Issue 1159 (www.techweb.com).
113. Fuller, D, Akinwande, A & Sodini, C (2003) 'Leading, Following or Cooked Goose? Innovation Successes and Failures in Taiwan's Electronics Industry', *Industry and Innovation,* Vol. 10, No. 2, p. 189.
114. Hung, SC (2002) 'The co-evolution of technologies and institutions: a comparison of Taiwanese hard disk drive and liquid crystal display industries', *R&D Management*, Vol. 32, No. 3, p. 187.
115. Fuller *et al.*, 2003:190.
116. This figure has been calculated by adding the market shares of Japan and South Korea together for 2003. South Korea's global market share stood at 44 per cent (Source: Samsung Electronics, Special Report, *The Economist,* 15 January 2005, p. 67). Japan's market share stood at 34 per cent (Source: 'Flat out for flat screens: the battle to dominate the $29 billion market is hotting up but the risk of glut is growing', *Financial Times,* 24 December 2003).
117. Linden *et al.*, 1998:22.
118. *Ibid*:26.
119. *Ibid.*
120. *Ibid.*
121. *Ibid.*
122. Linden *et al.*, 1997:19.
123. Linden *et al.*, 1998:26.
124. *Ibid.*
125. *Ibid.*
126. *Ibid.*
127. *Ibid.*
128. *Ibid.*
129. *Ibid.*
130. *Ibid.*
131. *Ibid*:26–27.

9

'Missing the Boat' in Liquid Crystal Displays: The Strategies of the US and Western Europe in a New Sectoral System of Innovation

This chapter analyses the 'failed' strategies pursued by the US and Western Europe to catch up in TFT-LCDs in the 1990s. Even though both strategies were ultimately unsuccessful, they are very different from one another.

It should be noted that during the 1990s two attempts were made in the US to enter the TFT-LCD industry. The first was little more than an ill-conceived attempt by a handful of small firms to halt the advances Japanese firms were making in the US TFT-LCD market. It was the second, led by the government, which was the most determined and coordinated effort to establish a significant foothold in the TFT-LCD industry. The government felt compelled to take the lead because it was all too aware of the industry's serious problems, and it feared this might have implications for national security. Government might have been aware of the industry's problems and had every intention of addressing them, but what the government never anticipated was that its strategy would be seriously affected by a 'problem' that the country has never accepted that it suffers from: it is poor at production. This is the principal factor that fatally undermined the US's strategy and eventually caused it to fail. It is also the same factor that led Western Europe's strategy to fail.

When it embarked on its strategy, Western Europe, quite unwittingly, fell into the same trap as the US. Like the US, it has never come truly to appreciate the importance of production, especially in electronics. As discussed in Chapter 4, Western Europe has long been the laggard in the semiconductor industry. Many of the problems which Western Europe has experienced in that industry are connected with its failure to master the art of production. If semiconductors have been 'difficult' for Western Europe, TFT-LCDs have proved far more problematic than it could ever have imagined.

United States

In Chapter 4, it was discussed how innovative firms, assisted by large military procurement during the 1950s and 1960s, combined to give the US an unassailable lead in semiconductors. The magnitude of the lead was such that it took Japan roughly two decades to catch up to a position where it could challenge the US. In Chapter 6, the analysis focused on how innovative firms once again allowed the US to develop an early lead, this time in LCDs, for it then to give the lead away to Japan because the firms which had been responsible for it, abandoned the very technologies which they themselves had pioneered. The actions of RCA and Westinghouse under very different circumstances may not have been quite so damaging and were potentially reversible. As shown in Chapter 7, however, the turmoil in the US digital watch industry, and the failure of US firms to cope with the switch to LCD technology and the concomitant financial pressures, put paid to this and only served to greatly exacerbate an already serious situation. As things stood, by the late 1980s the US LCD industry had become a 'cottage' industry.

It was at about this point that the US government, and computer firms in particular, belatedly recognised how important LCDs had become. Knowing it was vital that the country should not miss out on the huge opportunities TFT-LCDs offered it was the government that decided in 1994 to spearhead a strategy to catch up in the industry. The core of the government's catch-up strategy was the formation of a government–industry R&D consortia. Under the guidance of the Advanced Research Projects Agency (ARPA),[1] the research arm of the Department of Defense (DOD), the government formed with those firms still left in the LCD industry a consortium known as the United States Display Consortium (USDC). This is a superb example of the implementation of the much-vaunted 'new' US technology policy of the early 1990s, which was *expected* to greatly enhance the competitiveness of domestic industries.

The USDC was set up in July 1993, and its initial members were ARPA, AT&T, Xerox, Tektronix and nine other smaller firms.[2] The express objective was to, 'Develop and organise the manufacturing expertise to develop the infrastructure required to support a world-class

United States based manufacturing capability for high-definition flat panel displays'.[3] To achieve its objective, membership to the USDC was open to manufacturers of displays, manufacturers of display materials and equipment, and end-users. Through the pooling of resources and know-how it was hoped that this would reduce the risks to firms of developing new technologies, and also help reduce their R&D costs.[4] After much debate between the members of the consortium, it was decided that the USDC would concentrate on developing technologies needed for a wide variety of different flat panel display technologies. No single technology was going to be given priority at the expense of developing others.[5]

The USDC has close parallels with Sematech, which was set up during the 1980s for semiconductors.[6] It has not been as fruitful as Sematech, nor will it ever be. At least in semiconductors the US had an industry that it could take steps to protect while efforts were made to make it more competitive, but in LCDs there is no industry to speak of to protect. The basic idea of the USDC was to build an industry from scratch. In theory this seems quite a straightforward proposition, but in practice the idea suffers from an 'invisible' flaw that makes it almost unworkable. In Chapter 6, it was noted that when Peter Brody approached corporate America to invest in TFT technology it did not want to know. The problems that Brody experienced during the 1980s are symptomatic of a much deeper malaise that still affects the country today. In contrast to Japanese and South Korean firms which have long put production at the centre of their corporate strategies, US firms have given it a low priority for far too long and the country has paid a very heavy price for it. As competition from Japanese and South Korean firms has become more intense over the years, US firms have shown not only an increasing reluctance to compete with them head-on, but a distinct inability to do so. This is what has made its recent effort to catch up in TFT-LCDs an impossible task.

The legacy of disinvestment which firms actively engaged in from the late 1970s onwards has left permanent damage. With the haemorrhaging of knowledge and experience to produce TN-LCDs and STN-LCDs, along with the closure of so many production facilities, the chances of US firms of climbing aboard the TFT-LCD bandwagon and staying on it were very slim indeed. The damage done to the LCD industry hit the supporting industries equally badly. Thus, by the time those US firms still in the LCD industry contemplated joining the TFT-LCD bandwagon, they discovered the supporting infrastructure had all but disappeared, competition was intense, and investment outlays had soared to stratospheric levels. Only with government support did they feel confident enough to take substantial risks to establish any sort of real presence in a very dynamic industry.

The importance of production

The low priority that US firms have generally given to production probably stretches over a period of more than twenty years. The extent of the problem has been documented by Cohen & Zysman (1987).[7] They argue that since the early 1970s the common assumption made was that it did not matter if a significant proportion of manufacturing industry was moved offshore, or in some instances completely disappeared. The decline in the importance of manufacturing was seen as inevitable, and anyway it would be replaced in time by services and high-technology.

This view was contained in a report on trade agreements to the US Congress in the mid-1980s that offered a framework to understand why the country's trade position had deteriorated so badly. In the report it was stated that:

> The move from an industrial society toward a 'postindustrial' service economy has been one of the greatest changes to affect the developed world since the Industrial Revolution. The progress of an economy such as America's from agriculture to manufacturing to services is a natural change.[8]

As Cohen & Zysman point out, the transition from manufacturing to services has been confused with the transition from agriculture to manufacturing. Where the major difference lies between the two respective transitions is that the US did not get out of agriculture in the same way as it has done in manufacturing. The reason why the shift out of manufacturing should be taken so seriously, at least according to Cohen & Zysman, is what they have described as 'direct linkages'.[9] In short, once a country has lost its technological leadership it should expect a considerable contraction in the number of highly skilled, service-related jobs.[10] To back up their argument they cite the example of the steel industry. When the industry was flourishing, the country used to export engineering services in areas such as the design, construction and operation of steel mills. All that changed when the industry succumbed to competition from Japan and other competitors, and it started to import those services it had once exported.[11]

The seriousness of the country's production problem is reflected in the huge trade deficits. Between 1977–85 the accumulated trade deficits alone amounted to $505.6 billion.[12] The huge deficits were not just a feature of the early and mid-1980s. They continued throughout the rest of the 1980s and persisted throughout the 1990s and into the 2000s. The accumulated deficit for 1990–2005 is $4,687 trillion.[13] As the country's trade position failed to radically improve during the 1980s, another major investigation was launched to find out what was going wrong. This time round, the report rejected outright the previous explanation that it had been caused by some 'natural' economic transition. Instead, it identified a myriad of failings within US industry which if left unattended threatened to bring the country to its knees; short time horizons, weaknesses in development, outdated production strategies, industry and government not working together, and poor education and training. The report which identified these failings and caused a huge stir in the process was the study, *Made in America* (1989), carried out by the Massachusetts Institute of Technology (MIT) Commission on Industrial Productivity.[14] It started:

> To live well, a nation must produce well. In recent years many observers have charged that American industry is not producing as well as it ought to produce, or as well as it used to produce, or as well as the industries of some other nations have learned to produce. If the charges are true and if the trend cannot be reversed, then sooner or later the American standard of living must pay for the penalty.[15]

In the mid-1990s, after much of US industry had gone through a long period of extensive restructuring, it was claimed in a survey that those industries which had suffered a huge loss of competitiveness in the 1980s, like cars, machine tools and semiconductors had made a 'spectacular comeback'.[16] One of the main examples that has been widely used to substantiate this claim has been the resurgence of Intel. Its continued domination of the global microprocessor market allowed the firm, in 1994 alone, to make a $2.3 billion profit with a turnover of $11 billion.[17] Clearly, there are other examples which like Intel have made huge progress since the mid-1980s, but this so-called 'renaissance' is nowhere near as great as has been claimed. The survey points out one key area where this renaissance has not been seen: consumer electronics. As mentioned in the MIT report, the short-time horizons of US firms were among the prime failings identified and these have genuinely had disastrous consequences. As Japanese firms poured investment into the industry to build up a lead, US consumer electronics firms, in contrast, diversified into areas such as car rentals and financial services in the pursuit of higher profits, which eventually led to the complete demise of the US industry.[18] The report continues, 'The famous contrast here is the way that America's RCA squandered its technological lead in liquid-crystal displays while Japan's Sharp worked doggedly to make brilliant use of the technology'.[19]

LCDs

The first attempt made in the US to enter the TFT-LCD industry came in mid-1990 when a small number of firms formed a new association known as the Advanced Display Manufacturers of America (ADMA).[20] It filed an anti-dumping petition with the US International Trade Commission (ITC) and the Department of Commerce, in which it was alleged some of the major Japanese electronics firms were guilty of 'predatory pricing', selling displays in the US at well below cost price.[21] Among the key 'evidence' submitted to support the petition was an interview given by an executive of Toshiba to the *Japan Economic Journal*, in which it was quoted Toshiba was 'prepared to accept red ink for the first five or six years' to establish itself in TFT-LCDs.[22]

A year after the petition was filed the ITC voted to authorise the imposition of anti-dumping duties on TFT-LCDs, electroluminescent displays, and sub-assemblies manufactured in Japan. The duties imposed ranged from 7.02 per cent for electroluminescent displays to 62.67 per cent duty for TFT-LCDs.[23] The reason for this huge variation in the range of duties imposed was because the ITC recognised there are several very distinct types of flat panel displays, whereas ADMA had wanted the ITC to treat it as a single industry.[24] After much wrangling between ADMA on one side, and by the Japanese firms and US computer firms on the other, the ITC made a judgement in favour of ADMA. During its investigation the ITC found on the basis of the evidence submitted, and what it had collected, there was a strong 'correlation' between the small size of the domestic industry and Japanese firms' pricing policies.

In view of the very small size of the industry, it seems the US firms were doing no more than wasting their time filing a petition they hoped would seriously dent the competitiveness of Japanese firms, which in reality stood no chance of success. It is important to realise, however, at the time when the petition was filed, the firms also had the explicit support of the ARPA and other government departments.[25] The government had become concerned in the late 1980s at the absence of any major domestic suppliers of flat panel displays for military use. To correct this serious situation, in December 1988, ARPA issued a Broad Area Announcement (BAA) inviting firms to submit proposals for research grants to develop flat panel displays. Two of the firms awarded contracts for research on LCDs that later became members of ADMA, were Optical Imaging Systems (OIS) and Magnascreen.[26] Once the research contracts had been awarded, ARPA then initiated a series of meetings to foster informal networks between the various firms as a way of strengthening the industry, and finding out what problems firms were experiencing. Through these meetings the main complaint the US firms had was the tendency of some of the Japanese firms to engage in predatory pricing. For the US firms concerned, this was having a damaging effect on them. Apart from putting their margins under pressure, it was making venture capitalists very reluctant to invest capital in future development work, as they had one very legitimate fear – these small firms were in no real position to compete with their much larger Japanese competitors.[27] Another serious complaint made came from a firm that found it impossible to win supply contracts from the domestic computer firms. Every time the firm competed for a contract it was undercut by a Japanese competitor. This made it difficult for the firm to build additional production capacity and compete for other large contracts.[28] With these sorts of problems, some of the firms made a collective decision to form ADMA, and file an anti-dumping petition as a way out of their difficulties. It was not all that long before their actions came back to haunt them.

When the ITC announced its decision in ADMA's favour, it created a furore in the US, as well as in Japan. It is understandable that the Japanese firms must have felt aggrieved, but so too did the US computer firms. When the anti-dumping investigation was carried out, it was made clear to the ITC that all the US computer firms were highly dependent on TFT-LCDs from Japan. Without a reliable source the production of notebooks in the US would rapidly grind to a halt. The chief counsel for Apple Computer informed the ITC that, 'There was simply no US manufacturer of active-matrix LCDs capable of supplying the quantities we needed'.[29] The computer firms such as Apple were only interested in one thing, their supplies of TFT-LCDs. Earlier efforts made by ADMA to convince the computer firms, and other fellow members of the American Electronics Association (AEA), of the 'strategic' importance of flat panel displays, had fallen on deaf ears.[30] There was consternation among the computer firms about the imposition of the anti-dumping duties, which led several of the big firms to announce they would be moving the production of notebooks off-shore. For example, Apple said it would transfer production to either Ireland or Singapore, and Toshiba said that all production would be transferred back to Japan.[31] As the duties were not applicable to the final product, the computer firms remained free to make notebooks in third countries, and then export them to the US without incurring any additional duties. In the end, the biggest loser from the anti-dumping petition was not the Japanese firms it had been aimed at, but the US, with the loss of a considerable number of jobs and the creation of serious splits within the US electronics industry.

After the fiasco of the ADMA anti-dumping petition a decision was taken in April 1994 to make a second attempt to enter the LCD industry. This time though it was done in a much more coordinated manner with the objective of improving the competitiveness of US firms, rather than resorting to protectionism. The Clinton administration announced the launch of its Flat Panel Display Initiative (FPDI), a project to be coordinated through ARPA, which was to invest $578 million over a five year period in flat panel display technologies.[32] When the FPDI was launched it was not explicitly spelled out that it was the government's objective to help completely rebuild the domestic LCD industry. However, a careful look at the words used by Kenneth Flamm (1994), who was the Principal Deputy Assistant of Defense, and Special Assistant to the Deputy Secretary of Defense at the time, indicates that it was.

Reflecting the government's concern about the plight of the flat panel display industry and the chances of its recovery, he states that, 'there is little production capacity in this country for mainstream products and little is likely to be built absent some new initiative. Japanese companies hold 95 per cent of the market'.[33] He has obviously chosen his words very carefully, but the mere mention of the lack of productive capacity, and Japan's share of the global market is a dead give-away. In this case the words 'flat panel display' is synonymous with TFT-LCD. Actual confirmation that the government was seriously intent on intervening in the LCD industry surfaced in an article in *The Economist*. It cited a DOD discussion document that had been leaked to the press, which mentioned that $500 million had apparently been earmarked for the construction of four TFT-LCD production facilities in various locations around the US.[34]

The broad aim of ARPA was to help create a vibrant industry through a series of measures aimed at encouraging the existing firms left in the industry to engage in greater collaboration and make investment more attractive. The *Financial Times* stated that this represented the largest ever US government funded programme for commercial technology.[35] Through the FPDI, ARPA hoped to increase the country's share of the global market from less than 3 per cent to around 15 per cent by the year 2000.[36]

The government felt prompted to try a second time because its earlier fears about national security had not gone away, if anything they had become more intensified. Flamm claimed that urgent measures were needed to reverse the dangerous 'pathological state'[37] of the domestic industry. As he explains:

> Our national security demands that US military forces have guaranteed, cost-effective access to the world's best technology. One technology area in which the Department of Defense (DOD) has serious concerns is flat-panel displays. These modern alternatives to the cathode-ray tube (CRT) will be needed ubiquitously, in everything from fighter planes, helicopters, and tanks to units of command posts. And they will make possible a vast new array of superior military products. Flat-panel displays are listed on virtually every critical technology list, yet DOD has no assured source for this technology – not domestic technology, captive DOD suppliers, or even overseas vendors.[38]

The DOD's fear of lack of guaranteed supplies was not totally unfounded. Sharp had made it quite clear to the DOD that it was not interested in either becoming a major supplier or making customised products for it.[39] Another case arose in 1992 when Japanese firms rationed supplies of LCDs to Taiwanese firms. The shortage of supplies meant they were unable to fulfil a large number of orders for notebooks.[40] Flamm points out that another obstacle to Japanese firms supplying the DOD were Japan's own military-export controls. The dual-use export policy lacked clarity which was an issue the government was unwilling to address.[41] While these fears were very real, the major worry in the US about the FPDI was the objective of creating substantial industrial capacity. For Flamm, this was of fundamental importance, as he recognised that US firms faced too many barriers to entry which in the past had put them off building a major production facility.[42] The recognition by the government of the industry's structural weaknesses, and the limitations of what the government itself could do, opened the way for the participation of foreign firms in the FPDI as this was one of the ways of assisting domestic firms overcome the entry barriers facing them.

The FPDI was structured in such a way that it would only support the 'pre-competitive' aspects of development. Capacity and manufacturing expertise would be developed through a 'dual-use' strategy. For the DOD this marked a new 'revolutionary' change to its procurement policy. In the past it tended to rely on firms that have supplied only military markets. Now the DOD wanted to use firms that supplied both military and commercial markets. Because the pace of technological change is far too rapid for the DOD to cope with on its own, it would instead provide assistance to develop technology aimed at the commercial market and then adapt it as appropriate. This would in theory allow the DOD access to leading edge technologies, thus avoiding the need to 'pick a winner' in an industry where there are several alternatives to choose from.

The FPDI comprised three major constituents: to increase investment in core R&D programmes and infrastructure, form R&D partnerships with firms committed to production, and stimulate demand. The first two elements were sound objectives, whilst the latter was of a more dubious nature. Increasing investment in R&D and infrastructure is certainly nothing new, but the establishment of a manufacturing test-bed by the DOD would allow firms to undertake important work before committing themselves to full-scale production. Without sufficient experience of the techniques and processes used, US firms would be at a permanent disadvantage, always destined to play the role of catch-up. By tying R&D funds to actual production, Flamm believed this was the best way of building up capacity. He did not want the DOD to repeat one of the big mistakes it had made in the past, simply giving funds to firms to carry out R&D without any strings attached. Between 1989–94 ARPA had already spent $300 million on R&D into flat panel displays, but this had not been sufficient to encourage firms to invest in any real capacity. Firms that wanted DOD funds were subject to stringent criteria. They were obliged to build capacity from their own resources; fund half the cost of any project supported by the DOD; be prepared to work to its requirements; and devise projects that comply to tight specifications.

The government's objective of encouraging the building of productive capacity was certainly not shared by everyone. A big debate was sparked off about whether it constituted a fundamental step towards a more comprehensive industrial policy, which is generally looked upon in the country with contempt. One of the fiercest critics of the FPDI was Barfield (1995).[43] His first criticism is of the idea that $578 million is considered by the government as being enough to get the industry off its knees. While he had no fundamental objections to the participation of overseas firms in the FPDI, he points out that about $3 billion of investment would be nearer the mark if the industry were to capture 15 per cent of the global market by 2000. As this investment has not been forthcoming in the past, he casts doubt on why it should in the future.[44] Japanese firms have been free to build as many production facilities as they wish, but they have chosen not to.

A second criticism that Barfield had of the FPDI was the DOD's aim of stimulating demand. Nothing at all is gained from artificially stimulating demand, especially for a product for which *actual* demand was predicted to grow at a healthy rate for several years to come. Flamm claims the DOD was sure it could use procurement as a way of driving new applications and boosting product sales by offering financial inducements to firms keen to integrate new display technology into prototypes of military systems.[45] Barfield, on the other hand, claims this was utter stupidity as the government was aware of the intentions of Japanese and South Korean firms to massively expand their production capacities; and to cap it all the DOD would account at most for 5 per cent of domestic demand.[46] Another weakness related to the building of capacity was that different production lines would have to be installed in any facility for those TFT-LCDs produced for military and civilian purposes. TFT-LCDs are delicate at the best of times, and those to be used in military applications would have to be built to completely different specifications from those used for civilian purposes if they are to withstand severe shocks, extreme changes in temperature and rugged conditions.[47]

The last relevant criticism of Barfield's is whether this constitutes a more fundamental move towards an industrial policy. Unlike Flamm who has not openly said that the FPDI was an attempt by government to use fears about national security as a cover to rebuild an civilian industry, Barfield argues this is a very good example of a government doing what a government should not be doing, attempting to 'pick winners'. He notes that the idea for the FPDI originally came from the White House and not the DOD. The main figures involved were the former Head of the National Economic Council, Laura Tyson, who is known to favour 'cautious activism', and two of her former colleagues, Robert Rubin and Bowman Cutter.[48]

When the FPDI was launched, Barfield alleges that White House officials made it clear that the real goal was to boost the competitiveness of the civilian industry. Cutter is meant to have gone even further in an interview, and said it was to create 'a model of technological development that will equip US companies to break into markets already seized by the Japanese', implying this was the beginning of a new 'era' to foster several new civilian technologies that had already been identified.[49] Seizing on this, Barfield used it as his evidence to accuse the government of trying to 'pick winners', which could only end in failure. He argued that the FPDI would fail because 'the state of the technology is still in a state of flux and researchers are exploring a number of different approaches and materials'.[50] The point he has 'conveniently' missed is that with the FPDI, the government was trying to help a domestic industry exploit a technology that is dominated by firms from East Asia. In light of the overwhelming evidence how can the government be wrong in this instance?

In a reply to the criticisms made by Barfield, the two issues which Flamm (1995) answers directly are the notion that the White House instigated the FPDI and that it was part of some new 'industrial policy' initiative.[51] As one might expect from a person who was working for the DOD, Flamm argued that the idea for the FPDI came from the DOD. What Flamm 'forgets' to mention though, is that it was only possible to get the FPDI launched after President Clinton had been elected, and individuals like Laura Tyson and Robert Rubin were brought into government to put some different ideas into practice. As far as the latter point is concerned, Flamm observes that since 1989 the DOD had already invested a considerable amount of money in R&D on flat panel displays, but that this had not led to any firms building any productive capacity. It was of no benefit to the government if valuable R&D funds were not being used in the way that had originally been intended. The obvious thing to do was to tie future R&D funds to production, which is a reasonable precondition for those firms wanting government R&D funds. The government recognised that to make the FPDI work it was important for all the parties involved to keep to their side of the agreement.

Ignoring these rather superficial issues, Flamm and Barfield have good arguments both for and against the FPDI. On the one hand, Flamm was correct to say the government was right to be worried about a domestic industry that is on its knees and needs assistance if it is to have a hope of recovering and competing at a global level. The US cannot afford simply to sit back and let a key technology with huge commercial potential slip through its hands. On the other, Barfield is on very firm ground when it comes to investment. The amount that ARPA proposed to spend over five years was barely enough to scratch the surface of the problem. As the industry evolves, the levels of investment do not go down, only up. What the FPDI does not do is solve private firms' lack of access to long-term capital at highly competitive rates of interest – something which Japanese firms do have the luxury of. This is reinforced by the example of IBM. In the mid-1990s, it was the only US firm producing TFT-LCDs for the civilian market, but its production facility is located in Japan.[52] The lack of sufficient capital for long-term investment was certainly a major factor in its decision to locate its joint venture with Toshiba in Japan. It also graphically highlights the huge task that much smaller firms face in obtaining the capital they require. If a firm like IBM has had problems in obtaining capital, the chances of smaller firms of obtaining what they need must be very remote. Barfield's other legitimate criticism is ARPA's idea to stimulate demand, which does not in the end help the government or the firms involved.

After analysing the various pros and cons of the FPDI, the litmus test of it surely is whether the country's share of the global market in TFT-LCDs has increased. Compared to the situation in the early 1990s, the position of the US has actually deteriorated quite substantially. In March 1999, it emerged that the last remaining military supplier of AM-LCDs in North America, dpiX, a subsidiary of Xerox, was on the verge of being sold off or shut down.[53] This came on top of a spate of similar closures. In 1997 Hyundai closed its ImageQuest AM-LCD subsidiary in Fremont, California, and Litton Industries shut down its AM-LCD operations in Canada.[54] Then, in September 1998, Guardian Industries Corporation closed down its AM-LCD operations, OIS, another former supplier to the military.[55]

Why have Litton Industries, Guardian Industries Corporation, and now Xerox decided that it is in their interests to dispose of their AM-LCD operations? The answer as ever appears to be financial. Looking at what dpiX experienced in 1998, the unmistakable impression is that in supplying the DOD with AM-LCDs, it was impossible to realise the vital economies of scale, and to generate the necessary stream of income required for future long-term investment. It has been estimated that in 1998 dpiX shipped more than $10 million worth of products, but at the same time suffered losses totalling $15 million.[56] The divestment of Litton Industries and Guardian Industries Corporation from AM-LCDs, and Xerox's 'inability' to supply TFT-LCDs to the DOD on a profitable basis, effectively brought the US government's efforts to foster a vibrant TFT-LCD industry to the end of the road. Evidence that the DOD had finally reached crisis point in terms of trying to source domestic supplies of TFT-LCDs came in June 2000. It emerged that the DOD had concluded a 10-year purchasing agreement with LG Philips LCD to buy up to $1 billion worth of TFT-LCDs for use in US military and commercial aircraft cockpits.[57] For the US government and US firms this is a disastrous state of affairs. If US firms cannot even maintain a significant presence in a captive market, then how can they be expected to survive in the highly competitive civilian market? Something is obviously not working very well.

Western Europe

In the 1970s the prospects for a prosperous LCD industry developing in Western Europe were not very good, even though there were a number of firms which could have formed the basis for one. It is noteworthy that the LCD sector is at variance with the majority of other major industrial sectors. In most major industrial sectors there is normally a significant European presence, but in LCDs there never really has been. The LCD industry in Western Europe failed properly to develop during the course of the 1970s for two principal reasons. Firstly, European electronics firms dismissed LCDs as nothing more than a passing fad, which would not last beyond the heyday of the digital watch,[58] and secondly, the intense competition from Japanese firms in the digital watch and calculator industries. Several of the European firms who were actively engaged in the development and production of LCDs retreated from the sector as the prospects for the industry in Western Europe went from bad to worse.

The main problem for Western Europe in the early 1970s was that there were no large firms, similar to Japan's Sharp, that were prepared to make a long-term commitment to developing LCD technology. The catalyst for this was the fierce price competition in the US digital watch industry. The intense Japanese competition was too much for many of the European firms to endure and they were subsequently forced to leave the LCD industry altogether. By 1983 those firms that had survived the turmoil in the digital watch industry were forced to rethink their strategies. In light of the prevailing circumstances, they chose to take the same route as the US firms had done, focus on serving niche markets, producing custom and specialised displays, where there was no Japanese competition.[59] Bahadur (1983) points out that European (and US) firms were already at a disadvantage to Japanese firms even at this very early stage of the industry's development. The perennial Japanese strengths in production and large-scale investment had already given them a competitive advantage over small and medium-sized producers in terms of quality and costs.[60]

The role of Philips

In hindsight, efforts to develop a 'proper' European LCD industry were very slow in getting off the ground. It was not until concerns about the plight of the consumer electronics industry began moving up the political agenda at the European level in the early to mid-1980s, and there was mounting evidence pointing to LCDs being on the verge of a period of explosive growth, that efforts to foster an indigenous industry really got under way. Philips assumed the primary responsibility of this huge task, in conjunction with a handful of close collaborative partners. As the last remaining major consumer electronics firm in Western Europe, the strategy Philips has followed to tackle the Japanese domination of LCDs has been vital in helping to determine the eventual outcome of its collaboration with other European firms, and the EU's research programmes for LCDs. For both Western Europe and Philips, the need for an indigenous industry could not be greater, nor the stakes higher. However, despite this the competitive challenge has proved too much, and the prospects for an indigenous industry are now bleaker than ever before. Exactly why Philips has 'failed' is due to a combination of factors. Some though, have been more instrumental than others in contributing to this conspicuous lack of success. Two of the problems that Philips has struggled with are old ones that have plagued the European consumer electronics industry since the 1970s, weaknesses in production and innovation. The firm's problems have been further compounded by the absence of major end-users and shifts in the demand for certain sized displays that have caused prices to plummet.

The strategy that Philips has used to establish a foothold in LCDs has to be seen in a wider perspective. Ever since the late 1970s, Philips has struggled to compete successfully against Japanese firms in important product areas, in spite of having considerable technological strengths in them. In theory any firm that is technologically strong in a product should be able to compete successfully on a reasonably level playing field, but Philips defies conventional wisdom and has become somewhat of a mystery. It was during the 1980s that Philips's inability to compete with Japanese firms suddenly came to the fore, when the demand for the video-cassette recorder (VCR) began to rise very rapidly throughout much of Western Europe. In the previous decade Philips had been in a very good position to exploit VCR technology. Although Japanese firms and US firms had been actively developing VCR technology during the 1950s and 1960s, it was Philips who was the first to introduce and launch a technically acceptable videotape recorder for the consumer market at a competitive price in 1972.[61]

Within little more than a decade that early lead had been completely whittled away, placing Philips in a very precarious position. As is widely known, it was competition from Japan that was responsible for this. The ensuing standards battle between Sony's Betamax system and JVC's VHS system, effectively made the V2000 system, which Philips had developed in cooperation with Grundig, completely worthless. Understandably, this was a disaster for Philips on an unimaginable scale. Following the problems with the VCR, Philips was radically restructured to make it more competitive and to eliminate the chances of a repetition of the VCR fiasco.

The worrying thing for Philips is that despite being restructured and its corporate strategy being given a thorough overhaul, it had not visibly improved its ability to compete against East Asian firms. The parallels between Philips's experience in VCRs and LCDs is much more than a straightforward coincidence. Somehow Philips has allowed itself to fall into the same trap. In view of the enormous loss of prestige and financial costs suffered by Philips with the VCR, one might have expected the firm to have avoided doing the same thing all over again. But surprisingly, that is precisely what it has done in LCDs. It developed its own technology, that is technically simpler than TFT technology, in the hope of competing successfully against Japanese firms in an area where they are strong. Since the early 1990s, the competitive pressures have steadily built up on Philips as the industry has become more dynamic, and the firm has found itself less and less able to keep pace with the leading firms. Faced with the prospect of being eventually left behind in the LCDs, Philips made an announcement in late 1996 that amounted to an admission that it had finally capitulated, in other words, the competitive challenge from Japan had once again proved too much for it. To gain a bigger share of the fast growing TFT-LCD market, Philips announced it was going to 'abandon' its own technology and would adopt Japanese technology.

This begs the obvious question: why has history repeated itself? This can only be properly answered by looking at Philips's corporate strategy since the 1980s and the principal factors that caused the V2000 system to fail. Philips's corporate strategy in the 1980s was very different from the one it pursued in the 1970s. One of the most fundamental changes made to it was the move away from the 'try and do everything' to a 'make and buy' policy. The former was the legacy of Philips's expansionist and diversification policy which it had pursued vigorously prior to the 1970s, but which by the 1980s had clearly become very outdated.

Fundamentally, the new 'make and buy' policy was forced on Philips because of its deteriorating financial position which reached crisis point in 1990, when it reported a record loss of £1.3 billion, an unprecedented figure in the firm's 99 year history.[62] These heavy losses forced the firm's management to concede that 'it can no longer to afford to be in the forefront of developing every type of technology. Nor can it continue to develop and build-in-house every significant component that it needs for its products'.[63] Under the new policy any technological expertise that Philips required would be bought in from outside. Savings made elsewhere would offset any large costs that Philips may incur as a consequence of this policy.[64] The other major change to Philips's strategy, designed to enhance its technological capabilities and overall competitiveness, was to increase the number of joint ventures and strategic alliances it had with other firms. Having cut back its own technological activities, this was the only path Philips could follow to successfully exploit the range of emerging technologies, especially digital technologies. This put Philips in a position whereby it could reduce the costs of R&D, share technology, and enter rapidly growing markets that may otherwise be closed off to it. To make its strategy a success Philips has concluded agreements with a host of firms covering a wide range of technologies.

At the same time Philips was overhauling its corporate strategy and setting out a new path to follow, it was also repeating some of the serious errors it had made in the past without realising what it was doing. With its eyes firmly fixed on the future, Philips was busy concentrating on sorting out what it was convinced were the firm's problem areas. During this process it somehow 'overlooked' key factors that were responsible for the failure of the V2000 system, severe delays in getting the product to the market and the lack of sufficient production capacity.

One of the first errors that Philips made in its bid to become a major producer of LCDs was its decision to develop its own technology at the time when the South Korean chaebols, and the majority of Japanese firms were either clambering or already aboard the TFT bandwagon. The technology that Philips developed is simpler than TFT technology. It is still an active-matrix display, but in place of TFTs are diodes. By developing this technology Philips had hoped to gain a major competitive advantage over its competitors, as a diode LCD can be produced at lower cost than a TFT-LCD because it requires fewer manufacturing steps. The other major attraction of the diode LCD for Philips was that its performance was comparable with the TFT-LCD. With the relative simplicity of diode technology and the performance it offered, Philips must have presumed the technology would be a winner.

With the continued Japanese domination of the industry, Philips rapidly discovered that the benefits it had expected to accrue from its diode technology did not materialise. The firm was caught completely by surprise when the diode LCD failed to make any serious headway against the TFT-LCD. Its conspicuous lack of success eventually forced Philips's hand, and in late 1996 the firm made the announcement that it was to adopt Japanese technology. According to a Philips's official at the time, Mr Joel Crolla, 'We [Philips] have a good system and we believe in it, but the problem is that we have to do everything ourselves',[65] so much for the 'revamped' corporate strategy. After nearly a decade Philips was finally going to get aboard the TFT bandwagon it had long assumed it could afford to ignore. The way it managed to climb aboard the TFT bandwagon was to develop a joint venture with Hosiden (called Philips-Hosiden FPD) and to take a 50 per cent stake in LG LCD.

This whole situation has been brought about by the failure of a joint venture that Philips formed in 1993, in partnership with Thomson and Sagem, as the European 'challenge' which was supposed to capture 8 per cent of the global market by 2000.[66] The three firms formed the Flat Panel Display Company, based in Eindhoven, a production facility with a capacity to produce around 75,000 panels per month. Of the three partners, Philips had the largest stake in the joint venture with 80 per cent, with Thomson and Sagem each having a 10 per cent stake. In 1994 Philips reduced its stake by selling 10 per cent of its share to the German chemical firm, Merck.[67] On the production side, Philips initially intended to use the diode technology once the facility was on stream and then later switch over to a technology co-developed with

Sagem.[68] Furthermore, between 25–30 per cent of output was to be used for internal purposes, with the rest exported to markets in Western Europe and North America, and only a small percentage going to markets in Asia.[69]

The joint venture was formed under the auspices of the EU's Framework Programme as an example of 'successful' European collaboration. The EU has been under a lot of pressure to set up ventures of this kind because one of the most persistent criticisms made of the Framework Programme is that it is a waste of European taxpayers' money. A substantial amount of money has been invested in the programme over the years, but in that time there have been relatively few tangible products developed. This is bad for a programme that was set up partially to assist the EU to compete with Japan, and without visible evidence of the development of commercially viable products, future programmes are at risk of being severely cut back or at worse completely scrapped. Having said that, as a large producer of cathode-ray tubes (CRTs), the move by Philips into LCDs made good commercial sense. It was not just a complementary investment, but if Philips was to have a long-term future in displays it had to move into the fast growing sector and gradually lessen its reliance on CRTs.

By the time, however, that the joint venture had commenced full production, its long-term viability was already looking increasingly uncertain. Philips took much longer than it should have done to get its own diode LCD on to the market. In 1987 Philips built a pilot production line to test the viability of its technology, but it was another four years before the decision was taken to go ahead with mass production. Once that decision had been taken there was a three-year gap before production finally got under way. The original start date for production was January 1993,[70] but this had to be postponed for a year due to production problems. Philips found it needed to have considerably more production experience than it had in LCDs to get yields up to acceptable levels. Precisely because of this lack of experience, yields at the start of commercial operations were approximately 10 per cent, which then slowly increased to approximately 40–50 per cent.[71] Yields at these levels meant that Philips lost considerable sums of money that it could ill afford. Notwithstanding that, the firm was faced with another serious problem, the LCD market was changing faster than it had previously thought possible.

Certainly, one of the main motives for Philips forming its joint venture with Hosiden appears to have been the collapse in the price of 10.4-inch displays, from which Philips expected to make considerable profits. Prices dropped rapidly because the supply of these displays far exceeded demand, and this was combined with a more general move towards 11.3-inch and 12.1-inch displays. In six months, between April–September 1995, the price for a 10.4-inch TFT-LCD fell from $1,045 to $500, and by March 1997, the price had not recovered, but declined even further to $450.[72] Philips could not afford to sustain even more losses than it had already done, nor did it have the flexibility to switch over to making larger displays where demand was outstripping supply. Hosiden, the other hand, had the capacity to make these larger displays. In April 1998, the Flat Panel Display Company was reportedly phasing out the production of displays based on Philips diode technology and reset LCD technology, and switching over to using Hosiden's amorphous silicon TFT technology.[73]

When the Flat Panel Display Company was set up, it was always Philips's intention to build another LCD production facility. As with the first one, Philips wanted collaborative partners who were willing to help share the burden of investment, and expected the attractiveness of its diode technology would make other firms eager to join up with it. The most likely location for the plant was somewhere in Asia as Western Europe suffers from two serious impediments that make it a very unattractive location for an investment of this nature.[74] The first one relates to the lack of major end-users of LCDs, that is, vibrant computer and consumer electronics industries. It is absolutely pointless for a firm to build a major production facility in a location where there are no major end-users, which are vital to maintain the health and vitality of the LCD industry.

The second concerns the supply of supporting technologies that are vital for the production of LCDs; they are simply not available in Western Europe. For example, the machines that deposit the semiconductor materials onto the glass substrate of a display. The only manufacturers who produce this type of machinery are in Japan. Another example is the specialised glass that the displays are made from. The glass has to withstand temperatures of up to 600°C, but this also has to be imported from Japan as none of Western Europe's glass manufacturers produce it.[75] It has been suggested that at least 30 per cent of these supporting technologies are unavailable in Western Europe.[76] The only area where Western Europe does excel is in the supply of liquid crystal chemicals. The leading manufacturer and supplier of these highly specialised chemicals is Merck. This on its own, however, is insufficient to make any location in Western Europe attractive enough for a second production facility, as Merck has a long established presence in the Japanese market, where it has steadily built up its operations since 1975 when restrictions on foreign investment were lifted.

Confronted with this array of problems, the adoption of Japanese technology was Philips's only means of salvation. Through its joint venture with Hosiden (then the smallest of the Japan's TFT-LCD firms) Philips derived two key benefits. Firstly, it established an important foothold in the Japanese LCD market, and secondly, it gained vital access to TFT and production technologies. For Hosiden, the main benefit from teaming up with Philips was the $270.3 million investment it received for its LCD operations. When the price of 10.4-inch displays collapsed the firm suffered disproportionately compared to the much larger Japanese firms. Like Philips, it suffered financial losses it could ill afford. As the TFT-LCD market became more competitive, Hosiden never had the opportunity to recover financially from this setback. Under these mounting financial and competitive pressures Hosiden was eventually forced to exit the industry and Philips bought Hosiden's remaining stake of the venture.[77] This cemented Philips's position in Japan's LCD market. Philips's position in TFT-LCDs was further enhanced with its stake in LG LCD. The main advantage Philips has gained from its stake in LG LCD is access to additional production capacity and expertise that was absolutely essential if was to increase its global market share.[78]

The European Union

Unlike the situation with VCRs, where the EU made a number of belated responses to a problem that was really beyond its control, in LCDs it has adopted a perceptibly 'hands-off' approach. As there is no indigenous industry to 'protect', the EU's only measurable contribution to the development of a European industry has been support through its ESPRIT programme. In the fourth programme (1994–98) which had an overall budget of $16 billion, just $128 million was allocated for display projects, with $118 million going to fund five LCD projects.[79] Of the $118 million, $108 million, was used to fund four linked AM-LCD projects, with the other $10 million funding one LCD ferroelectric project. The $108 million allocated for the four AM-LCD projects was distributed between the participants, which included 4 firms, 17 industries, 2 institutes and 3 universities in 6 countries. For the other project, $10 million was distributed between 5 industries and 2 universities in 4 countries. ESPRIT funded half the cost of all these projects.[80]

Another common criticism made of the Framework Programme is that all too often research funds are spread too thinly to be effective. From the above, it would seem that this was the case with LCDs. The experience of Philips, however, which has been a major participant in the EU's AM-LCD research programmes tells us that this is part of a much more fundamental competitiveness problem facing Western Europe. Certainly, the funding arrangements of the Framework Programme need to be reformed, but it is well known that simply spending more money on R&D is not always the solution to a problem. Now that the Framework Programme has been in existence for more than a decade, the EU has accepted the fact that research is only one of the ways needed to improve Western Europe's competitiveness. Having recognised the extent of the problem, in December 1995 it launched a draft Green Paper on Innovation in a renewed attempt at exploring a range of options to improve it.

In the document all of the old familiar problems are there: insufficient investment in R&D and industrial research, lack of adequate sources of finance; an unfavourable tax and regulatory environment; a failure to anticipate technological trends; lack of proper coordination between member states; and poorly adapted education and training systems.[81] It goes on to add that Western Europe continues to suffer from a 'paradox'. Over the past fifteen years its scientific performance has been excellent, but its technological and commercial performance in high technology sectors such as electronics and information technologies has deteriorated.[82] The EU's diagnosis of itself is a correct one, and on the surface it appears that no matter what it does it just cannot seem find the right formula to restore its competitiveness.

Even after all these years, however, the importance of production is not fully appreciated. It is barely given a mention in the Green Paper. All of this is quite surprising in light of the astonishing success of the chaebols and the problems Philips has long experienced in semiconductors,[83] and more recently, in TFT-LCDs. Until the EU, and European firms like Philips appreciate just how important it is to get the production side of their operations right, Western Europe's search for improved competitiveness will remain as elusive as it ever has been, and it will continue to lose out in TFT-LCDs.

Summary

In no other industrial sector has the cost to the US and Western Europe for being weak in production been so high. Their respective strategies to catch up in TFT-LCDs have unearthed a very serious production problem, which has not manifested itself fully until now. There is more than enough evidence to suggest that it has been lurking just below the surface for a very considerable period of time. For example, the huge US trade deficits and Philips's 'eternal' struggle to compete successfully in consumer electronics, indicates that something is not right. In the one industry where it was most likely to come out into the open, semiconductors, especially because of the very high rate of technological change, the US and Western Europe have drawn upon extensive production experience to maintain a substantial presence in the industry by improving their competitiveness.

With TFT-LCDs, this option has been closed off to them. It is not the lack of previous production experience in them that is the fundamental problem for the US and Western Europe. One only has to look at Taiwan to see that it is possible to build up a presence in TFT-LCDs, albeit slowly. Although Japan has supplied it with substantial amounts of TFT-LCD technology, it is only of any use to Taiwan if it can properly utilise it. Taiwan has shown it can make full use of the technology because its expertise in production is on a par with Japan's and South Korea's. Taiwan could simply not make progress in TFT-LCDs, compete in DRAMs, or consumer electronics, if its expertise in production did not run wide and deep. What Taiwan has, the US and Western Europe lack, depth and breadth in the production field as a whole, which the TFT-LCD industry demands. It is this that is primarily responsible for the US and Western Europe not catching up in TFT-LCDs. The firms have not properly learnt about the technology.

Notes

1. ARPA was formerly known as DARPA (Defense Advanced Research Projects Agency).
2. Hart, JA (1994) 'Policies Toward Advanced Display in the Clinton Administration' in *Advanced Flat Panel Display Technologies*, Proceedings of the International Society for Optical Engineering (SPIE), 7–8 February 1994, San Jose, California, Vol. 2174, p. 20. The other firms were Electro-Plasma, Kent Digital Signs, Norden Systems, Optical Imaging Systems (OIS), Photonics Imaging, Planar Systems, Silicon Video, Standish Industries, and Three Five System.
3. Quoted in Fennell, LE (1994) 'Flat Panel Display Manufacturing Infrastructure Development', in *Advanced Flat Panel Display Technologies*, Vol. 2174, p. 26. The focus of the paper by Fennell is on how the USDC was structured and organised.
4. Hart, 1994:21.
5. *Ibid.*

6. Sematech is otherwise known as the Semiconductor Manufacturing Technology Research Consortium and is jointly funded by private industry and the DOD. It includes 14 American semiconductor companies, among them IBM, AT&T and all of the major merchant suppliers. In addition, 80 per cent of the American equipment manufacturers are members. Membership for American firms is voluntary and depends on their willingness to contribute to the project's financing and to agree to its terms about licensing research results and the like (Tyson, L (1992) *Who's Bashing Whom: Trade Conflict in High-Technology Industries*, Washington, DC: Institute for International Economics, p. 152, footnote 79).

7. Cohen, S & Zysman, J (1987) *Manufacturing Matters: The Myth of the Post-Industrial Economy*, New York: Basic Books.

8. Quoted in *Ibid*:5.

9. *Ibid*:3.

10. *Ibid*.

11. *Ibid*.

12. Figure calculated from Table 5.1 in *Ibid*:63.

13. The figures for the years 1990 and 1991, were obtained from the *World Economic Outlook: International Monetary Fund*, May 1993, Table A28, p. 163. The figures for the years 1992 to 1995 were obtained from *The IMF Annual Report*, 1996, p. 54. The figure for the year 1996 was obtained from the *OECD Economic Outlook*, June 1997, Table: External Balances, p. 46. The figure for the year 1997 was obtained from the *OECD Economic Outlook*, June 1998, p. 43. The figure for the year 1998 was obtained from the *OECD Economic Outlook*, June 1999, Table: External Balances, p. 42, and the figure for 1999 was obtained from the *OECD Economic Outlook*, June 2000, Table: External Indicators, p. 54. The figures for 2000–2005 were obtained from the OECD *Economic Outlook*, June 2004, p. 44.

14. Dertouzos, ML, Lester, RK & Solow, RM (1989) *Made in America: Regaining the Productive Edge*, Cambridge, Massachusetts: MIT Press.

15. *Ibid*:1.

16. 'Nuts, Bolts, Chips: Look at us, we can make it', Survey of American Business, *The Economist*, 16 September 1995.

17. Intel's share of the global microprocessor market was 75 per cent (*Ibid*).

18. *Ibid*.

19. *Ibid*.

20. ADMA was formed by seven firms: Planar Systems, Optical Imaging Systems, Plasmaco, Cherry Corporation, Electro-Plasma, Photonics Technology and Magnascreen. Hart, JA (1993) 'The Anti-Dumping Petition of the Advanced Manufacturers of America: Origins and Consequences', *World Economy*, Vol. 16, No. 1, p. 85.

21. *Ibid*:87. The main Japanese firms the anti-dumping petition was aimed at were Matsushita, Sharp, Toshiba, and Hosiden.

22. Quoted in *Ibid*:101.

23. *Ibid*:103.

24. *Ibid*:102.

25. *Ibid*:101.

26. The two other firms which became members of ADMA and won ARPA contracts were Planar Systems and Photonics. The former for research on electroluminescent displays and the latter for plasma displays. *Ibid*:100.

27. *Ibid*:101.
28. *Ibid*.
29. Quoted in *Ibid*:104.
30. *Ibid*:101.
31. *Ibid*:105.
32. Flamm, KS (1994) 'Flat-Panel Displays: Catalysing a US Industry', *Issues in Science and Technology*, Fall, p. 27.
33. *Ibid*.
34. 'Flat Screens: Crystal Diplomacy', *The Economist*, 30 April 1994, pp. 84–87.
35. 'US to fund flat panel display', *Financial Times*, 28 April 1994.
36. Barfield, C (1995) 'Flat-Panel Displays: A Second Look', *Issues in Science and Technology*, Winter, p. 23.
37. Flamm, 1994:31.
38. *Ibid*:27.
39. *Ibid*:28.
40. Werner, K (1995) 'US Display Industry on the Edge', *IEEE Spectrum*, May, p. 62.
41. Flamm, 1994:28–29.
42. *Ibid*:29.
43. Barfield, 1995:25. The line adopted by Barfield is supported by Mowery, DC (1995) 'Flat-panel displays', *Issues in Science and Technology*, Spring, p. 5.
44. Barfield, 1995:23.
45. Flamm, 1994:30.
46. Barfield, 1995:23.
47. *Ibid*.
48. *Ibid*:22.
49. Quoted in *Ibid*.
50. *Ibid*.
51. Flamm, KS (1995) 'In Defense of the Flat Panel Display Initiative', *Issues in Science and Technology*, Spring, pp. 22–25.
52. Barfield, 1995:21.
53. 'Military reels as latest LCD maker falters', *Electronic Engineering Times*, 26 March 1999, p. 1. (www.eetimes.com).
54. *Ibid*:3.
55. *Ibid*.
56. *Ibid*:1.
57. 'LG Philips lands $1 billion deal to supply TFT-LCD panels to US', *Electronic Buyers' News*, 30 June 2000 (www.eoenabled.com/).
58. 'Europe's Liquid Crystal Assets – Philips is leading an assault on Japanese dominance of the active-matrix LCD market', *Financial Times*, 22 December 1994.
59. Bahadur, B (1983) 'A Brief Review of History, Present Status, Developments and Market Overview of Liquid Crystal Displays', *Molecular Crystals & Liquid Crystals*, Vol. 99, 350–51.
60. *Ibid*.
61. Dai, X (1996) *Corporate Strategy, Public Policy and New Technologies: Philips and the European Consumer Electronics Industry*, Oxford: Pergamon, p. 103.
62. *Ibid*:180.
63. Quoted in *Ibid*:175.
64. *Ibid*.

65. 'Philips in LCD Joint Venture', *Financial Times*, 12 October 1996.
66. 'Europe's liquid assets: Philips is leading an assault on Japanese dominance of the active LCD market', *Financial Times*, 22 December 1994.
67. 'Philips reduces stake in LCD Unit', *Financial Times*, 24 June 1994.
68. 'Philips joins French in plan to beat Japanese', *Financial Times*, 27 November 1992.
69. 'Cars of the future will bristle with liquid crystal display (LCD) navigation devices' – 'Screen development: European Research Intensifies', Review of Information Technology, *Financial Times*, 3 May 1995.
70. 'Philips joins French in plan to beat Japanese', *Financial Times*, 27 November 1992.
71. Interview.
72. 'LCDs Move to the Desktop', International Features, *Byte*, March 1997 (www.byte.com/art/9703/sec16/art/art8.htm).
73. Company Reports, *Displays Focus*, April 1998, p. 2 (Displays Focus is a trade newsletter for the UK Display Industry).
74. 'Europe's liquid assets: Philips is leading an assault on Japanese dominance of the active LCD market', *Financial Times*, 22 December 1994.
75. Review of Information Technology, *Financial Times*, 3 May 1995.
76. Interview.
77. 'Philips buys Hosiden stake in LCD joint venture', *Electronic Engineering Times*, 1 September 2000 (www.eetimes.com).
78. 'Philips targets screens for a dazzling future', *Financial Times*, 19 May 1999.
79. Werner, 1995:69.
80. *Ibid.*
81. *Green Paper on Innovation*, (Draft) December 1995, European Commission, pp. 24–37.
82. *Ibid*:5.
33. Philips has withdrawn from the mainstream SRAM market, and halted the development and production of DRAMs above 4 Mbytes (Dai, 1996:161, 231).

10
Conclusions

In Chapter 1, it was stated that semiconductors and TFT-LCDs have been at the centre of the growth of the information age. Without the technological advances made in these (and other related) industries the information age would not have evolved at the rate it has. The dynamism displayed by these two industries raised fundamental questions about what supported and powered these two high technology industries forward. Furthermore, it was noted the two industries have some very close links. Aside from being found in many of the same products, the countries responsible for the breathtaking developments in the semiconductor industry are exactly the same ones that now dominate the TFT-LCD industry, notably East Asian firms. A closer examination revealed a paradoxical situation. The US was the first country to develop an early technological lead in semiconductors and LCDs. In semiconductors it took full advantage of this early lead. In LCDs, however, it 'gave away' its lead to Japan. By the early 1990s Japan had developed a monopoly in TFT-LCDs. In an effort to capture a stake in this rapidly expanding industry, Western Europe, the US, South Korea and Taiwan each launched their own catch-up strategies in the 1990s. Two of the principle questions that needed to be addressed were: how had this situation materialised? And why were there such wide variations in the performances between different countries in TFT-LCDs?

To recap the situation once more, in an era of liberalisation and intense global competition, it is extremely rare to come across an industry of this nature, dominated by a single country. From a historical perspective, the stranglehold that Japan then had over the TFT-LCD industry seemed impenetrable. Since the early 1970s Japan has concentrated on building up its presence in the industry and developing LCD technology, unlike Western Europe and the US which did not start to make any serious efforts to develop LCD technology until the late 1980s.

For the first time, Western Europe and the US faced the process of industrial catch-up to have any chance of establishing a foothold in this rapidly expanding industry. The challenge that faced Western Europe and the US was an enormous one. Japan had a huge lead and to close the gap Western Europe and US were required to cover a lot of ground within a relatively short period of time. That was just to prevent the lead from growing any bigger than it already was. That said, South Korea and Taiwan were in the same position as Western Europe and the US. They too were required to cover a lot of ground to close the gap with Japan. The difference between South Korea and Taiwan on one hand, and Western Europe and US on the other, was that the former were no strangers to the catch-up process, the latter were. Based on what South Korea and Taiwan demonstrated that they were capable of in semiconductors, their prospects for catching up in TFT-LCDs were good. The chances of Western Europe and the US catching up, compared to those of South Korea, were much less promising given the difficulties that they have both experienced in semiconductors. As it has turned out, South Korea and Taiwan met the challenge of catching up with Japan in TFT-LCDs very well, unlike Western Europe and US which, in essence, never got off the starting blocks.

Key to understanding why South Korea and Taiwan could catch up and the efforts of Western Europe and US were severely hindered, are their respective national systems of innovation (NSIs). Using Patel & Pavitt's (1994) system of classifying NSIs,[1] it is now possible to categorise each of these respective countries' NSI. In view of what we have seen in TFT-LCDs, it would seem appropriate to put Japan, South Korea and Taiwan in the 'dynamic' category, and Western Europe and the US in the 'myopic' category.[2] The factor that really sets the two systems apart from each other, as mentioned in Chapter 2, is how they treat technological activities.[3] In 'myopic' systems, investments in technological activities are treated in the same way as conventional investments. When decisions are made about conventional investments they are based predominantly on short-term considerations (Hayes & Abernathy, 1980). Because of their very nature, investments in technological activities involve accepting a high degree of risk and uncertainty. However, when they are evaluated using the same criteria as conventional investments there is a great tendency for them to appear unattractive. The structure of this type of NSI makes them *passive* learning systems, where learning is a highly uneven process.

In 'dynamic' systems, investments in technological activities are treated completely differently from conventional investments.[4] When investments are made in technological activities, they are made with the long term very much in mind. Risk, market uncertainty, and high levels of investment are all seen as being the norm. This makes it much easier to follow certain technological trajectories than would otherwise be the case. Investments made in early generations of a technology are often rapidly followed up by subsequent programmes of investment as the technology starts to mature. This ensures the process of learning goes uninterrupted and helps develop vital technological competencies.[5] The structure of this type of NSI makes them *active* learning systems, which facilitates a 'smooth' learning process.

From what we can ascertain from the above (and preceding chapters), how successful a country eventually becomes in a particular technology appears ultimately to hinge on how its firms manage to exploit it. As the firm is, so to speak, at the heart of the NSI, what firms do can either have a positive or negative impact on the NSI. In the great majority of cases, when a firm acts in what could be construed as a 'negative' way, for example, by halting work on a promising technology, the overall impact on an NSI is going to be negligible. The story is very different though when firms act in concert with each other and take the view that a particular technology is unimportant. On the odd occasion when this occurs, the effect on the performance of an NSI can be quite devastating.

The impressive performances of Japan, South Korea, and Taiwan in TFT-LCDs suggests that their NSIs are radically different from the NSIs of Western Europe and US. Whilst in some respects this is undoubtedly the case, there is one distinguishing feature which sets the Japanese, South Korean, and Taiwanese firms apart from their European and US counterparts, which is all too easy to miss. Although it is not part of the NSI itself, it does have a huge influence on its efficiency. Quite simply, Japanese, South Korean, and Taiwanese firms have been continuously active in LCDs from the first generation. This has enabled the firms to engage in continuous learning about the intricacies of the technology to develop important technological competencies in the field. Western European and US firms, on the other hand, stopped learning about LCDs in the 1970s when they dismissed them as being 'unimportant', and hence they now know far less about them.

It is argued here that as the LCD has gradually become more complex, the NSIs of Japan, South Korea, and Taiwan possessed the technological capabilities to successfully support its development, giving the firms the opportunity to develop their own competencies. Because Japan was the first country to make a long-term commitment to the LCD, its NSI was naturally the first to gear itself up for LCD development. With the successful transition from one generation to the next, Japan gradually increased the barriers to entry to the LCD industry. The three main barriers it had 'erected' by the time it had reached the third generation related to the technological complexity of the TFT-LCD, the difficulties of its production, and the investment outlays required to build production facilities. Japan might have been way ahead of anyone else in the industry in the early 1990s, but South Korea could mount a successful catch-up campaign because its NSI was already well structured for the development of highly complex and capital intensive technologies. By effectively utilising the same infrastructure it has used to excel in DRAMs, South Korea was able to overcome the barriers to entry with which Japan had ring fenced the LCD industry. Taiwan, on the other hand, had to find an alternative path.

Unlike the NSIs of South Korea and Taiwan, which coped with the pressures of catching up, the NSIs of Western Europe and the US fared very badly. The root cause of the problem has to do with the firms' lack of a continuous presence in LCDs. When the decisions were taken to catch up with Japan, the NSIs had no time to adjust to the demands that would be placed upon them. With no time to adjust, the NSIs from the very outset stood no chance of overcoming the barriers to entry, hence their very poor performance. Rothwell & Zegveld (1985) note that the EU is ill equipped to catch up as it suffers from a number of important weaknesses. In particular, Western European firms generally lack the capacity to excel in highly efficient, high-quality production, and they also lack the commercial dynamism and ability to capitalise on Western Europe's scientific creativity.[6] At least in TFT-LCDs, the most problematic entry barrier for Western Europe and US is perhaps the firm itself and its distinct lack of commitment to learning.

Another aspect related to the NSI is how globalisation has affected the development of specific technologies. Freeman & Soete (1997) contend that globalisation will have little impact on NSIs as the transnational corporations (TNCs), particularly those of Japan and South Korea, still conduct an overwhelming proportion of their technological activities at home.[7] As this has long been the main source of their strength they are not likely to do anything to damage it. This certainly seems to be the case in semiconductors where Japanese firms have consolidated amongst themselves and South Korea continues to dominate the DRAM sector. In TFT-LCDs, the NSI has been absolutely critical to the success of Japan, South Korea and Taiwan in the industry. It confirms beyond doubt Porter's (1990) point that a firm's national base is more important than ever before in an era of globalisation. Understanding why there have been huge variations in the performances of South Korea, Western Europe, and the US in TFT-LCDs explains only part of the industry's idiosyncratic development.

One final aspect related to the NSI are the two key questions Malerba raises about new sectoral systems of innovation. The first is how do new sectoral systems emerge? In the case of LCDs, it emerged as a very distinct technology in its own right, which was successfully commercialised in the 1970s with the development of the calculator and digital watch. At this point in time it had no obvious linkages with any other technology or sector. All that changed with the development of the third generation LCD, the TFT-LCD. Due to the nature of its complexity, the countries that wanted to develop it found it required a strong presence in one other very specific industry, the DRAM industry. The technologies to fabricate a TFT-LCD and DRAM were merging. So in the current context, it could be said that the TFT-LCD has 'emerged' from the DRAM sector. Apart from dominating the TFT-LCD industry, Taiwan and South Korea also have a very strong presence in the DRAM industry. Japan too has traditionally enjoyed a strong presence in the industry. Compare this to Western Europe and the US, which have no presence in the TFT-LCD industry to speak of, and are very weak in the DRAM industry. The conclusion that can be drawn from this is that by relinquishing a position in a technologically dynamic industry such as DRAMs, a country can end up forsaking future opportunities quite unwittingly, should a similar technologically dynamic industry emerge at a later date. The US is a country that has done precisely this. During the 1980s it relinquished its position in the DRAM industry, and when it tried to enter the TFT-LCD industry it found its attempts were severely constrained, amongst other things, by the chronic lack of production expertise in this type of industry. Much the same could be said about Western Europe.

The second question of Malerba's concerns the link with previous sectoral systems. In this instance this requires no more elaboration. The TFT-LCD industry has strong past and present links with the DRAM industry, but these links are by no means apparent when the two industries are analysed at a superficial level. In cases concerning other technologies, this is a vital question that needs to be asked. Links between different sectors are rarely immediately self-evident, but as the case of TFT-LCDs and DRAMs illustrates, developments in one sector can radically affect another.

Related to the link between TFT-LCDs and DRAMs is the role of users that have helped drive the TFT-LCD industry forward. In Chapter 4, it was explained that users such as the military and computer industry helped spur the growth and development of the US semiconductor industry, and the consumer electronics and computer industry did likewise for the Japanese semiconductor industry. The role of the computer and consumer electronics industries have been equally important in spurring the growth and development of the TFT-LCD industry. With Japan, South Korea, and Taiwan all having strong consumer electronics and computer industries, their TFT-LCD industries have managed to feed off these industries and vice-versa.

At this late stage it is very difficult to assess to what extent the growth of the TFT-LCD industry could have been constrained had Japan, South Korea, and Taiwan not had dynamic computer and consumer electronics industries of their own. Again, we only have to look at the semiconductor industry to see what the 'lack' of domestic users can have on a technologically dynamic industry. One of the main factors which 'stunted' the growth of Western Europe's semiconductor industry was the lack of a vibrant indigenous computer industry, which was dominated by US firms, notably IBM. The dominance of US firms of the European computer industry did nothing to help advance the European semiconductor industry, which is one of the reasons why it has never kept pace with its US and Japanese rivals.

The other facet of the industry's development that needed explanation was how Japan came to dominate the industry in the first place. For this, path dependency has been used to trace the path that the industry has followed from its earliest origins in the late nineteenth century through to the 1990s. With this approach a number of 'historical accidents' have been identified, that is, specific events that have had a disproportionate impact on the LCD industry and helped shape it into the industry it is today. Two of the most important ones Westinghouse and RCA decisions to abandon TFT technology independently of each other, and the turmoil in the US digital watch industry in the 1970s. Having unearthed the reasons why Japan came to dominate the industry in the way that it did, and why South Korea and Taiwan, and not Western Europe and US have caught up in TFT-LCDs, the development of the LCD industry is not the 'mystery' it originally was.

In the final analysis, with the TFT-LCD now established as the dominant flat-panel display (FPD) technology, the medium and long-term prospects for the industry look very promising. Future growth looks assured as the TFT-LCD starts to make inroads into areas where the cathode-ray tube (CRT) has long reigned supreme, notably in televisions. In this one sector alone, demand is huge but until now has remained largely untapped. When combined with continued growth in products such as notebooks, demand for TFT-LCDs can only continue to increase at a healthy pace for the industry. In light of this it is easy to see why Freeman & Soete (1997) say that the sectoral systems of innovation approach is one of the most exciting areas of future research in the millennium. Apart from being able to couple it with the NSI, new technologies are continuously emerging and it is impossible to predict their future path and how they may affect different sectors. In an increasingly competitive global environment, the urgency to gain a greater understanding of inter-linkages between sectors can only increase. Firms and countries alike are always on the look out for new ways to gain a competitive advantage. Knowing what is emerging over the horizon before your rivals do has never been more important, and what is more, the need to be first is never going to go away.

Notes

1. Patel, P & Pavitt, K (1994) 'National Innovation Systems: Why They Are Important, And How They Might Be Measured and Compared', *Economics of Innovation and New Technology*, Vol. 3, pp. 77–95.
2. *Ibid*:91.
3. *Ibid.*
4. *Ibid.*
5. *Ibid.*
6. Rothwell, R & Zegveld, W (1985) *Reindustrialisation and Technology*, Harlow: Longman, p. 256.
7. Freeman, C & Soete, L (1997) *The Economics of Industrial Innovation*, London: Pinter, pp. 307–309.

Bibliography

Abramovitz, M (1986) 'Catching up, Forging Ahead, and Falling Behind', *Journal of Economic History*, Vol. 47, No. 2, pp. 385–406.

Advanced Flat Panel Display Technologies, Proceedings of the International Society for Optical Engineering (SPIE), 7–8 February 1994, San Jose, California, Vol. 2174.

Amsden, A (1989) *Asia's Next Giant: South Korea and Late Industrialisation*, Oxford: Oxford University Press.

Anchordoguy, M (1989) *Computers Inc: Japan's Challenge to IBM*, Cambridge, Massachusetts: Harvard University Press.

Archibugi D, Michie, J & Howells, J (eds) (1999) *Innovation Policy in a Global Economy*, Cambridge: Cambridge University Press.

Archibugi, D & Mitchie, J (eds) (1997) *Technology, Globalisation, and Economic Performance*, Cambridge: Cambridge University Press.

Arthur, WB (1989) 'Competing Technologies, Increasing Returns and Lock-In by Historical Events', *Economic Journal*, Vol. 99, pp. 116–131.

Bahadur, B (1983) 'A Brief Review of History, Present Status, Developments and Market Overview of Liquid Crystal Displays', *Molecular Crystals & Liquid Crystals*, Vol. 99, pp. 345–374.

Bain, JS (1956) *Barriers to New Competition: Their Character and Consequences in Manufacturing Industries*, Cambridge, Massachusetts: Harvard University Press.

Barfield, C (1995) 'Flat Panel Displays: A Second Look', *Issues in Science and Technology*, Winter, pp. 21–25.

Besen, SM & Farrell, J (1994) 'Choosing How to Compete: Strategies and Tactics in Standardisation', *Journal of Economic Perspectives*, Vol. 8, No. 2, pp. 117–131.

Berggren, C & Nomura, M (1997) *The Resilience of Corporate Japan: New Competitive Strategies and Personnel Practices*, London: Paul Chapman.

Bond, J & Levenson, DM (1993) 'The US gears up to challenge Japan in flat panel displays', *Solid State Technology*, December, pp. 37–43.

Bork, RH (1978) *The Antitrust Paradox: A Policy at War with Itself*, New York: Basic Books.

Borrus, M & Hart, JM (1994) 'Display's the Thing: The Real Stakes in the Conflict over High Resolution Displays', *Journal of Policy Analysis and Management*, Vol. 13, No. 1, pp. 21–54.

Borrus, M, Millstein, JE & Zysman, J (1983) 'Trade and Development in the Semiconductor Industry: Japanese Challenge and American Response' in L Tyson & J Zysman (eds) *American Industry in International Competition, Government Policies and Corporate Strategies*, New York: Cornell University Press.

Branscomb, LM (1993) *Empowering Technology: Implementing a US Strategy*, Massachusetts: MIT Press.

Braun, E & MacDonald, S (1980) *Revolution in Miniature: The History and Impact of Semiconductor Electronics*, Cambridge: Cambridge University Press.

Brody, TP (1996) 'The birth and early childhood of active matrix: A personal memoir', *Journal for the Society of Information Display*, Vol. 4, No. 3, pp. 113–127.

Brody, TP (1984) 'The Thin-Film Transistor: A Late Flowering Bloom', *IEEE Transactions on Electron Devices*, Vol. ED-31, No. 11, pp. 1614–1628.

Bowonder, B, Sarnot, SL, Srinivas Rao, M & Poornachander Rao, D (1994) 'Electronic Display Technologies: State of the Art', *Electronics Information and Planning*, Vol. 21, No. 12, September, pp. 683–743.

Bowonder, B, Sarnot, SL, Rao, MS & Miyake, T (1996) 'Competition in the global electronics display industry: strategies of major players', *International Journal of Technology Management*, Vol. 12, Nos. 5/6, pp. 551–576.

Buigues, P, Jacquemin, AP & Sapir, A (eds) (1995) *European Policies on Competition, Trade and Industry: Conflict and Complementarities*, Aldershot: Edward Elgar.

Business Week (1994) 'Japan's Liquid Crystal Gold Rush', 17 January, pp. 44–45.

Byte (1997) 'LCDs Move to the Desktop', International Features, March.

Callon, S (1995) *Divided Sun: MITI and the Breakdown of High-Tech Industrial Policy 1975–1993*, Stanford, California: Stanford University Press.

Cantwell, J & Iammarino, S (2003) *Multinational Corporations and European Regional Systems of Innovation*, London: Routledge.

Castellano, JA (1988) 'Liquid Crystal Display Applications: The First Hundred Years', *Molecular Crystals & Liquid Crystals*, Vol. 165, pp. 389–403.

Castellano, JA (1970) 'Now the heat is off, liquid crystals can show their colors everywhere', *Electronics International*, 6 July, pp. 64–70.

Chen, CF & Sewell, G (1996) 'Strategies in technological development in South Korea and Taiwan: the case of semiconductors', *Research Policy*, Vol. 25, No. 5, pp. 759–785.

Cohen, S & Zysman, J (1987) *Manufacturing Matters: The Myth of the Post-Industrial Economy*, New York: Basic Books.

Christensen, JL (1992) 'The Role of Finance in National Systems of Innovation' in BA Lundvall (ed.) *National Systems of Innovations: Towards a Theory of Innovation and Interactive Learning*, London: Pinter, pp. 146–167.

Collings, P (1990) *Liquid Crystals: Nature's Delicate Phase of Matter*, Bristol: Adam Hilger.

Cooper, AC & Schendel, D (1988) 'Strategic Responses to Technological Threats', in ML Tushman & WL Moore (eds) *Readings in the Management of Innovation*, New York: Harper Business, pp. 249–258.

Cross, M & Hecht, J (1993) 'Fuzzy future for flat screens', *New Scientist*, 31 July, pp. 34–38.

Dai, X (1996) *Corporate Strategy, Public Policy and New Technologies: Philips and the European Consumer Electronics Industry*, Oxford: Pergamon.

Dalum, B, Holmen, M, Jacobssen, S, Preast, M, Rickne, A & Villumsen, G (1999) 'Changing the regional system of innovation', in J Fagerberg, P Guerrieri & B Verspagen (eds) *The Economic Challenge for Europe*, Cheltenham: Edward Elgar, pp. 175–200.

Dasgupta, P & Stiglitz, J (1988) 'Potential Competition, Actual Competition and Economic Welfare', *European Economic Review*, Vol. 32, pp. 569–577.

David, PA (1993) 'Path-dependence and predictability in dynamic systems with local network externalities: a paradigm for historical economics' in D Foray & C Freeman (eds) *Technology and The Wealth of Nations: The Dynamics of Constructed Advantage*, London: Pinter, pp. 208–31.

David, PA (1986) 'Understanding the Economics of QWERTY: The Necessity of History' in WN Parker (ed.) *Economic History and the Modern Economist*, Oxford: Basil Blackwell, pp. 30–49.

David, PA (1985) 'Clio and the Economics of QWERTY', *American Economic Review*, Vol. 75, No. 2, May, pp. 332–37.

Davidson-Frame, J & Narin, F (1990) 'The United States, Japan and the changing technological balance', *Research Policy*, Vol. 19, No. 5, pp. 447–455.

Davidson, JH (2002) *The Committed Enterprise: How to Make Vision and Values Work*, Oxford: Butterworth Heinemann.

Davis, WS & McCormack, AM (1979) *The Information Age*, Reading, Massachusetts: Addison-Wesley.

Dempa (1997) 'Nine domestic LCD makers to produce more than 2 trillion yen ($15.87 billion) worth of LCDs in fiscal 2000', 9 April.

Dempa (1996) 'Korean firms trying to catch up with Japanese TFT-LCD makers', 31 August.

Demsetz, H (1982) 'Barriers to Entry', *American Economic Review*, Vol. 72, No. 1, pp. 47–57.

Depp, SW & Howard, WE (1993) 'Flat Panel Displays', *Scientific American*, Vol. 226, No. 3, pp. 40–45.

Dertouzos, ML, Lester, RK & Solow, RM (1989) *Made in America: Regaining the Productive Edge*, Cambridge, Massachusetts: MIT Press.

De Wit, B & Meyer, R (1998) *Strategy: Process Content, Context*, London: Thomson Learning, 2nd edition.

Displays Focus (1998) Company Reports, April.

Edquist, C (1997) 'Systems of Innovation Approaches – Their Emergence and Characteristics', in C Edquist (ed.) *Systems of Innovation; Technologies, Institutions and Organisations*, London: Pinter, pp. 1–35.

Electronic Buyers' Guide (1999) 'Quanta close to forging FPD Deal', 10 May, Issue: 1159 (www.techweb.com).

Electronic Buyers' News (2000) 'Philips LG land $1 billion deal to supply TFT-LCD panels to US', 30 June.

Electronic Buyers' News (1999) 'Quanta, Sharp in tech-license/joint venture deal', 31 May, Issue: 1162.

Electronic Buyers' News (1999) 'Quanta close to forging FPD deal', 10 May, Issue: 1159.

Electronic Buyers' News (1999) 'Fujitsu, Chi-Mei in AM-LCD pact', 15 March, Issue: 1151.

Electronic Buyers' News (1998) 'Freeze frame – Korean LCD makers put expansion plans on hold', 26 January.

Electronic Engineering Times (2005) 'Hynix to prevail in WTO ruling over EU says report', 16 March.

Electronic Engineering Times (2004) 'Hitachi, Toshiba, Matsushita join on LCD TV panels', 31 August.

Electronic Engineering Times (2004) 'Joint Sony, Samsung LCD factory opens', 19 July.

Electronic Engineering Times (2003) 'AU Optronics teams with Fujitsu on LCD development', 28 January.

Electronic Engineering Times (2001) 'Japanese display makers strain to regain footing', 19 October.

Electronic Engineering Times (2000), 'Philips buys Hosiden stake in LCD joint venture', 1 September.

Electronic Engineering Times (1999) 'Military reels as latest LCD falters', 26 March.

Electronic Engineering Times (1996) 'Korean vendors mount LCD challenge to Japanese', 10 November, Issue: 980.

Electronics International (1994) 'Europe gets its first FPD fab', 12 December, p. 4.

Electronics International (1994) 'FPDs vs CRTs; When will the battle begin?', 12 December, p. 5.

Electronics International (1994) 'OIS, Apple ink AMLCD development deal', 13 June, p. 1.

Electronics International (1994) 'Sharp shows largest TFT-LCD panel', 13 June, p. 1.

Electronics International (1994) 'Korean TFT-LCD makers are ready to challenge Japan', 23 May, p. 6.

Electronics International (1994) 'DoD rides to US FPD industry's rescue', 9 May, p. 14.

Electronics International (1994) 'Japan, LCDs dominate FPD market', 11 April, p. 6.

Electronics International (1994) 'South Korea eyes TFT-LCD mass production milestone', 11 April, p. 6.

Electronics International (1994) 'EU alliance hopes to crack LCD market', 11 April, p. 8.

Electronics International (1994) 'Taiwan jumps on LCD bandwagon', 11 April, p. 9.

Electronics International (1994) 'Japanese LCD makers fight to remain dominant as market expands', 11April, p. 7.

Electronics International (1994) 'LCD market brightens Japan's gloomy '94 outlook', 10 January, p. 8.

Electronics International (1993) 'Samsung doubles LCD production', 22 November, p. 8.

Electronics International (1993) 'Korean makers call for government help', 27 September, p. 6.

Electronics International (1993) 'Korea's electronics giants face off in TFT-LCD market', 23 August, p. 1.

Electronics International (1993) 'EC clears Philips/Thomson/Sagem LCD joint venture', 10 May, p. 1.

Electronics International (1993) 'Hyundai's subsidiary to develop LCD technology', 10 May, p. 2.

Electronics International (1993) 'ARPA backs US LCD venture', 10 May 1993, p. 2.

Electronics International (1992) 'European consortium to attack active-matrix LCD market', 14 December, p. 3.

Electronics International (1992) 'Can the US put together a flat-panel consortium', 15 June, p. 40.

Electronics International (1992) 'Displays: The race for better yield changes LCD horizons', 15 June, p. 38.

248 *Bibliography*

Electronics International (1988) Electronics newsletter: 'Now, a Japanese active-matrix LCD that measures 14.26 in...', October, p. 17.

Electronics International (1987) International newsletter: 'Japanese hone color LCDs for laptop PCs by selling tiny TVs', 15 October, p. 54.

Electronics International (1986) International newsletter: 'The Japanese are pushing on bigger LCD panels', 16 October, p. 50.

Electronics International (1983) 'Japan shows off its flat-panel work', 6 October, pp. 101–102.

Electronics International (1983) 'Japanese flat panels flatten competition', 28 July, pp. 48–49.

Electronics International (1983) 'The US pioneers have fled the frontier', 28 July, p. 48.

Electronics International (1983) 'Color TV set with LCD screen fits in pocket', 31 May, pp. 85–86.

Electronics International (1983) New Products: 'Large LCD suits portable computers', 31 May, p. 170.

Electronics International (1981) 'Big makers bow out of watch LCDs', 8 September, pp. 50–52.

Electronics International (1980) 'Motorola bows out of LCD business', 28 February, pp. 48–50.

Electronics International (1978) Electronics newsletter: 'New product line of big LCDs due from Beckman', 2 February, p. 33.

Electronics International (1977) 'Japan's watch market calm', 8 December, p. 78.

Electronics International (1977) 'LCD startup bedevils watch firms', 18 August, pp. 74–75.

Electronics International (1977) Electronics newsletter: 'Digital watch shake-out to continue', 4 August, p. 25.

Electronics International (1977) 'Times goes on the offensive', 23 June, p. 102.

Electronics International (1977) 'Optel comeback based on displays', 26 May, pp. 52–54.

Electronics International (1977) 'Liquid crystals for watches are turning out to be in good supply', 28 April, p. 30.

Electronics International (1977) 'Gas-discharge-display sales soar', 17 February, pp. 65–66.

Electronics International (1977) 'Watch surge generates LCD shortage', 6 January, pp. 67–69.

Electronics International (1976) Electronics newsletter: 'Motorola decides not to supply microprocessor... and gets back into watch module business', 25 November, p. 35.

Electronics International (1976) Electronics newsletter: 'LCD display shortage is already here says Beckman', 9 December, p. 25.

Electronics International (1975) 'Electronics adds wristwatch frills', 7 August, pp. 46–48.

Electronics International (1975) 'Timepiece uses diode, LCD displays', 26 June, p. 31.

Electronics International (1975) Electronics newsletter: 'Fairchild seeks liquid-crystal firm', 1 May, pp. 25–26.

Electronics International (1975) International newsletter: 'Production boosts in Japan threaten calculator makers', 6 March, p. 47.

Electronics International (1975) 'Watch market: is 40 a crowd?', 20 February, pp. 34–35.

Electronics International (1974) International newsletter: 'Japanese to export $165 LCD digital quartz wristwatch', 14 November, p. 56.

Electronics International (1974) 'TI defers its watch entry', 31 October, pp. 29–30.

Electronics International (1974) 'Bowmar introduces digital LED watch', August, p. 40.

Electronics International (1973) 'Displays', 25 October, pp. 104–105.

Electronics International (1973) 'Liquid-crystal array is large enough to show CRT-type information', 25 October, pp. 29–30.

Electronics International (1973) 'To a reed-switch manufacturer, liquid-crystal displays come naturally', 25 October, p. 34.

Electronics International (1973) Electronics newsletter: 'Japan accelerates calculator exports', 16 August, p. 29.

Electronics International (1973) 'Field-effect LCDs may give watches push in the market', 16 August, p. 33.

Electronics International (1972) 'A case for new watchmakers', 22 May, pp. 59–62.

Electronics International (1971) International newsletter: 'Japan seeks curb on exports to US', 5 July, p. 4E.

Electronics International (1971) Electronics newsletter: 'RCA liquid crystals in watch display', 5 July, p. 17.

Electronics International (1971) 'US fires first shot at Japanese calculator lead', 15 February, pp. 37–38.

Electronics International (1970) 'New IC market: electronic watches', 21 December, pp. 83–84.

Electronics International (1970) 'US firms gird for calculator battle', 23 November, pp. 83–86.

Electronics International (1963) 'Displays: $200 Million', 29 March, p. 15.

Electronics International (1963) 'Large Displays: Military Market Now, Civilian Next', 25 January, pp. 24–26.

Electronics International (1962) 'Our Growing Markets: An analysis of the present and a look into the future', 5 January, pp. 41–72.

Electronics International (1961) *Coordination of Information on Current Research and Development in the Field of Electronics*, 20 September.

European Commission, *Green Paper on Innovation* (Draft) December 1995.

Farrell, J & Saloner, G (1986) 'Installed Base and Compatibility: Innovation, Product Preannouncements, and Predation', *American Economic Review*, Vol. 76, No. 5, pp. 940–955.

Farrell, J & Saloner, G (1985) 'Standardisation, compatibility, and innovation', *RAND Journal of Economics*, Vol. 16, No. 1, Spring, pp. 70–84.

Fennell, LE (1994) 'Flat panel display manufacturing infrastructure development', in *Advanced Flat Panel Display Technologies*, Proceedings of the International Society for Optical Engineering (SPIE), 7–8 February, San Jose, California, pp. 25–30.

Financial Times (2005) 'Electronic bugs cause recall of 1.3 million cars by Mercedes', 1 April.

Financial Times (2005) 'Hynix fails to hit profit forecast on probe provision', 4 March

Financial Times (2005) 'DaimlerChrysler: Jurgen Schrempp Interview', 4 March.

Financial Times (2005) 'Firing Ms Fiorina: Her successor should consider breaking up the company', 10 February.

Financial Times (2003) 'Flat out for flat screens: the battle to dominate the $29 billion market is hotting up but the risk of glut is growing', 24 December.

Financial Times (2003) 'Hynix condemns US move to impose tariffs', 19 June.

Financial Times (2002) 'Pressure builds on Seoul over Hynix: Creditors are contemplating a third multi-billion dollar bailout of the troubled chipmaker amid mounting protest', 9 December (USA Edition).

Financial Times (2002) 'South Korea denies Hynix subsidy claims', 23 November.

Financial Times (2002) 'Fujitsu, AMD in flash memory talks', 9 October.

Financial Times (2002) 'Chipmakers find benefits in cooperation', 26 September.

Financial Times (2002) 'Micron ends hopes of Hynix merger', 3 May.

Financial Times (2002) 'South Korea maintains foreign focus: Resistance to a Hynix/Micron deal has opened old wounds', 24 April.

Financial Times (2002) 'The chips are down, but hope still flickers. In spite of the recent travails of the world semiconductor industry, innovation is flourishing as chip suppliers invest in wireless data and networking technologies', 17 April.

Financial Times (2002) 'EU probes South Korean chipmakers', 26 July.

Financial Times (1999) 'Philips targets screens for a dazzling future', 19 May.

Financial Times (1998) 'Texas Instruments quits memory chips', 19 June.

Financial Times (1998) Survey of semiconductors: 'Risk factors for producers: A perilous business', 4 February.

Financial Times (1997) 'Perfect images on display', 11 December.

Financial Times (1997) 'Anti-dumping duties lifted in Korean D-Rams', 1 December.

Financial Times (1997) 'Chips are down as EU acts on dumping', 1 April.

Financial Times (1997) 'Brussels stalls on chip duties', 10 March.

Financial Times (1996) 'Philips in LCD Joint Venture', 12 October.

Financial Times (1996) 'Sharp and Sony link in flat TV screen venture', 20 September.

Financial Times (1996) 'US and Japan in agreement on chips', 3 August.

Financial Times (1996) 'US and Japan close to deal on microchips', 2 August.

Financial Times (1996) 'Brussels steps up chip pact pressure', 31 July.

Financial Times (1996) 'US and Japan seek 11th hour deal on semiconductor trade', 30 July.

Financial Times (1996) 'Japan takes tougher line in chips dispute with US', 1 July.

Financial Times (1996) 'Japan chip market more open', 18 June.

Financial Times (1996) 'Japanese split on chip pact demand', 13 June.

Financial Times (1996) 'Japan backs EU role in chip negotiations', 4 June.

Financial Times (1996) 'Tokyo rejects three-way semiconductor accord', 24 April.

Financial Times (1996) 'EU protest over chip pact', 12 March.

Financial Times (1996) 'Japan braced for pressure on US access', 5 February.

Financial Times (1995) 'Washington and Tokyo split on renewal of the US-Japan. semiconductor agreement – Kantor calls for market share pact to be reviewed. No need say the Japanese, it has already worked', 23 November.

Financial Times (1995) 'Cars of the future will bristle with liquid crystal display (LCD) navigation devices' – 'Screen development: European Research Intensifies', Review of Information Technology, 3 May.

Financial Times (1994) 'Europe's liquid assets', 22 December.

Financial Times (1994) 'The future is crystal clear', 8 November.

Financial Times (1994) 'Foreign chip sales take 21.9% of Japan's market', 15 September.

Financial Times (1994) 'Philips reduces stake in LCD unit', 24 June.

Financial Times (1994) 'US to plug holes in flat panel industry', 29 April.

Financial Times (1994) 'US to fund flat panel displays', 28 April.

Financial Times (1994) 'Japan "trapped" by chips import deal', 23 March.

Financial Times (1994) 'Jump in foreign chips sales', 19 March.

Financial Times (1994) 'Tokyo "backsliding" on chips accord – Trade relations in jeopardy, US manufacturers warn Clinton', 9 February.

Financial Times (1994) 'Screen Technologies: Too soon to write off the cathode ray tube', Mobile Computing Survey, 26 January.

Financial Times (1993) 'A flat vision of the future', 10 September.

Financial Times (1993) 'How to stand out in a crowd', 10 September.

Financial Times (1992) 'Philips joins French in plan to beat Japanese', 27 November.

Flamm, KS (1995) 'In Defense of the Flat-Panel Display Industry', *Issues in Science and Technology*, Spring, pp. 22–25.

Flamm, K (1994) 'Flat-Panel Displays: Catalysing a US Industry' *Issues in Science and Technology*, Fall, pp. 27–32.

Florida, R & Browdy, D (1991) 'The Invention that Got Away', *Technology Review*, Vol. 94, Part 6, pp. 43–54.

Forester, T (1993) *Silicon Samurai: How Japan Conquered the World's IT Industry*, Oxford: Blackwell.

Foster, RN (1988) 'Timing Technological Transitions' in ML Tushman & WL Moore (eds), *Readings in the Management of Innovation*, New York: Harper Business, 2nd edition, pp. 215–228.

Fox, B (1996) 'Tektronix technology fuels Sony's flat screen programme' *International Broadcasting*, January, pp. 13–14.

Fransman, M (1990) *The Market and Beyond: Cooperation and Competition in Information Technology in the Japanese System*, Cambridge: Cambridge University Press.

Freeman, C (2002) 'Continental, national and sub-national innovation systems-complementarity and economic growth', *Research Policy*, Vol. 31, No. 2, pp. 191–211.

Freeman, C (1997) 'The "national system of innovation" in historical perspective', in D Archibugin & J Michie (eds) *Technology, Globalisation and Economic Performance*, Cambridge: Cambridge University Press, pp. 24–50.

Freeman, C (1988) 'Japan: a new national system of innovation?' in G Dosi *et al. Technical Change and Economic Theory*, London: Pinter, pp. 330–349.

Freeman, C (1987) *Technology Policy and Economic Policy: Lessons from Japan*, London: Pinter.

Freeman, C (1982) *The Economics of Industrial Innovation*, London: Pinter, 2nd edition.

Freeman, C & Perez, C (1988) 'Structural crises of adjustment, business cycles and investment behaviour' in G Dosi *et al.* (eds) *Technical Change and Economic Theory*, London: Pinter, pp. 458–479.

Freeman, C & Soete, L (1997) *The Economics of Industrial Innovation*, London: Pinter, 3rd edition.

Fuller, D, Akinwande, A & Sodini, C (2003) 'Leading, Following or Cooked Goose? Innovation Successes and Failures in Taiwan's Electronics Industry', *Industry and Innovation*, Vol. 10, No. 2, pp. 179–196.

Galli, R & Teubal, M (1997) 'Paradigmatic Shifts in National Innovation Systems', in C Edquist (ed.) *Systems of Innovation; Technologies, Institutions and Organisations*, London: Pinter, pp. 324–370.

Gilchrist, J & Deacon, D (1990) 'Curbing subsidies' in P Montagnon (ed.), *European Competition Policy*, London: Pinter, pp. 31–51.

Graham, MBW (1986) *RCA and the VideoDisc: The Business of Research*, Cambridge: Cambridge University Press.

Gulick, G (1992) 'Liquid crystal displays: the game changes', *Information Display*, Vol. 8, No. 12, December, pp. 11–13.

Hamel, G & Prahalad, CK (1994) *Competing for the Future*, Boston, Massachusetts: Harvard University Press.

Hart, JA (1994) 'Policies toward advanced display in the Clinton Administration' in *Advanced Flat Panel Display Technologies*, Proceedings of the International Society for Optical Engineering (SPIE), 7–8 February, San Jose, California, Vol. 2174, pp. 16–25.

Hart, JA (1993) 'The Anti-Dumping Petition of the Advanced Manufacturers of America: Origins and Consequences', *World Economy*, Vol. 16, No. 2, pp. 85–111.

Hayes, F (1991) 'Displays – Down the Dram Drain?', *Byte*, February, pp. 232–33.

Hayes, RH, Wheelwright, SC & Clark, KB (1988) *Dynamic Manufacturing: Creating the Learning Organisation*, New York: The Free Press.

Hayes, RH & Wheelwright, SC (1984) *Restoring Our Competitive Edge: Competing Through Manufacturing*, New York: John Wiley & Sons.

Hayes, RH & Abernathy, WJ (1980) 'Managing Our Way to Economic Decline', *Harvard Business Review*, July–August, pp. 67–77.

Heilmeier, GH (1976) 'Liquid Crystal Displays: An Experiment in Interdisciplinary Research that Worked', *IEEE Transactions on Electron Devices*, Vol. ED-23, No. 7 July, pp. 780–88.

Hirsch, P & Gillespie, J (2001) 'Unpacking Path Dependence: Differential Valuations Accorded History Across Disciplines' in R Garud & P Karnoe (eds) *Path Dependence and Creation*, Mahwah, New Jersey: Lawrence Erlbaum Associates Publishers.

Hitt, MA, Dacin, M, Tyler, B & Park, D (1997) 'Understanding the Differences in Korean and US Executives' Strategic Orientations', *Strategic Management Journal*, Vol. 18, No. 2, 159–167.

Hobday, M (1995) *Innovation in East Asia: The Challenge to Japan*, Aldershot: Edward Elgar.

Holmes, P (1993) 'Competition, Trade and Technology Policy' in M Humbert (ed.) *The Impact of Globalisation on Europe's Industries and Firms*, London: Pinter, pp. 96–105.

Holzler, H (1990) 'Merger control', in P Montagnon (ed.) *European Competition Policy*, London: Pinter, pp. 9–30.

Howells, J (1999) 'Regional Systems of Innovation', in D Archibugi, J Howells & J Michie (eds), *Innovation Policy in a Global Economy*, Cambridge: Cambridge University Press, pp. 67–93.

Innovation and Technology Transfer (1994) 'ESPRIT: Flat Panel Displays: A European Challenge', DG XIII/IV, European Commission, Vol. 15, No. 2, April, p. 21.

Hung, SC (2002) 'The co-evolution of technologies and institutions: a comparison of Taiwanese hard disk drive and liquid crystal display industries', *R&D Management*, Vol. 32, No. 3, pp. 179–190.

International Herald Tribune (2001) 'In Blur of Text and Image, Hybrid Industry Emerges', 18 April.

International Herald Tribune (1997) 'LG Semicon Seeks to Play Down Memory Chips', 14 March.

International Herald Tribune (1996) 'Sharp Takes a Lead in Flat Panel Displays', 13 August.

International Herald Tribune (1995) 'Big Screens Lure Japan Firms', 28 June.

International Herald Tribune (1995) 'A Flat Market for Flat Panel Displays', 30 May.

International Herald Tribune (1994) 'Koreans, Again, Challenge Japan: Bid to Break Hold on Flat-Panel Computer Screens', 13 May.

Johnson, G & Scholes, K (2002) *Exploring Corporate Strategy*, Financial Times, Harlow: Prentice Hall, 6th edition.

Johnson, K (1992) 'Flat panel displays or bust?', *Physics World*, Vol. 5, No. 9, September.

Jorde, TM & Teece, DJ (1992) 'Innovation, Cooperation and Antitrust' in TM Jorde & DJ Teece (eds) *Antitrust, Innovation and Competitiveness*, New York: Oxford University Press.

Johnson, C (1982) *MITI and the Japanese Miracle: The Growth of Industrial Policy 1925–1975*, Stanford, California: Stanford University Press.

Journal of Electronics Industries (1996) Korean Report: 'Makers Dodge Impact of Panel Glut. For Samsung 22 Inch Panel is Sales Elixir', June, p. 50.

Journal of Electronics Industries (1996) Korean Report: 'For Money, Technology, Korean TFT-LCD Makers Knit Alliances', June 1996, p. 50.

Kagaku Kogyo (1996) 'Hyundai Electronics starts TFT colour LCD panel production', 7 October.

Kaiser, R & Prange, H (2004) 'The Reconfiguration of National Innovation Systems – the example of German biotechnology', *Research Policy*, Vol. 33, No. 3, pp. 395–408.

Kahaner, D (1994) Summary of key papers presented at the Electronic Display Forum, Yokohama, Japan, 6–8 April, US Office of Naval Research, Tokyo, Japan.

Kay, J (1993) *The Foundations of Corporate Success: How Business Strategies Add Value*, Oxford: Oxford University Press.

Kelker, H (1973) 'History of Liquid Crystals', *Molecular Crystals & Liquid Crystals*, Vol. 21, pp. 1–48.

Kim, L (1997) *Imitation to Innovation: The Dynamics of Korea's Technological Learning*, Boston, Massachusetts: Harvard Business School Press.

Kim, L (1993) 'National System of Industrial Innovation: Dynamics of Capability Building in Korea', in RR Nelson (ed.) *National Innovation Systems: A Comparative Study*, New York: Oxford University Press, pp. 357–83.

Kim, L & Dahlman, CJ (1992) 'Technology policy for industrialisation: An integrative framework and Korea's experience', *Research Policy*, Vol. 21, No. 5, pp. 437–452.

Kim, SR (1996) *The Evolution of Governance and the Growth Dynamics of the Korean Semiconductor Industry*, Working Paper No. 20, Sussex European Institute, University of Sussex.

Korea Economic Daily (1997) 'Korean TFT-LCD Makers Most Likely to Take Lead Over Japanese', 10 August.

Korea Economic Weekly (1997) 'Korea TFT-LCD Output Will Rise Sharply Next Year', 11 August.

Korea Times (2000) 'Samsung, LG Philips Rank 1st, 2nd in TFT-LCD Market', 17 February.

Korea Times (1999) 'LG Jacks Up Production Capacity to Meet Soaring Demand for TFT-LCD', Business section, 4 August.

Korea Times (1999) 'Korean Chip Shares 41 Percent of World Market', 14 March.

Korea Times (1999) 'Hyundai Emerges as World's Second Largest DRAM Maker', 12 March.

Korea Times (1999) 'Samsung, LG Strengthen TFT-LCD Production Facilities', 18 February.

Korea Times (1999) 'Korean Firms Leading TFT-LCD Market', 18 February.

Korea Times (1999) 'TFT-LCD Makers to Claim 35 Percent of World Market', 11 January.

Krugman, P (ed.) (1986) *Strategic Trade Theory and the New International Economics*, Cambridge, Massachusetts: MIT Press.

Levin, RC (1978) 'Technical Change, Barriers to Entry and Market Structure', *Economica*, Vol. 45, pp. 347–361.

Linden, G, Hart, J, Lenway, S & Murtha, TP (1998) 'Flying Geese as Moving Targets: Are Korea and Taiwan Catching Up With Japan in Advanced Displays', *Industry and Innovation*, Vol. 5, No. 1, pp. 11–34.

Linden, G, Hart, J & Lenway, S (1997) 'Advanced Displays in Korea and Taiwan', *BRIE Working Paper 109*, December, University of California.

Link, A (1998) 'The US Display Consortium: Analysis of a Public/Private Partnership', *Industry and Innovation*, Vol. 5, No. 1, pp. 35–50.

Liu, SJ & Lee, JF (1997) 'Liquid Crystal Display Industry in Taiwan', *International Journal of Technology Management*, Vol. 13, No. 2, pp. 308–325.

Lundvall, BA (ed.) (1992) *National Systems of Innovation: Towards a Theory of Innovation and Interactive Learning*, London: Pinter.

Lundvall, BA, Johnson, B, Sloth Anderson, E & Dalum, B (2002) 'National systems of production, innovation and competence building', *Research Policy*, Vol. 31, No. 2, pp. 213–231.

Lynch, R (2003) *Corporate Strategy*, Financial Times Prentice Hall: Harlow, 3rd edition.

Macher, JT, Mowery, DC & Hodges, DA (1998) 'Reversal of Fortune? Recovery of the US semiconductor industry', *California Management Review*, Fall 1998, Vol. 41, No. 1, pp. 107–136.

Malerba, F (ed.) (2004) *Sectoral Systems of Innovation: Concepts, Issues and Analysis of Six Major Sectors in Europe*, Cambridge: Cambridge University Press.

Malerba, F (2002) 'Sectoral Systems of Innovation and Production', *Research Policy*, Vol. 31, No. 2, pp. 247–264.

Malerba, F (1985) *The Semiconductor Business: The Economics of Rapid Growth and Decline*, London: Pinter.

Markides, C (2000) *Making All the Right Moves: A Guide to Crafting Breakthrough Strategy*, Boston, Massachusetts: Harvard Business School.

Mathews, JA (1998) 'Fashioning a new Korean model out of the crisis: the rebuilding of institutional capabilities', *Cambridge Journal of Economics*, Vol. 22, pp. 747–759.

Mathews, JA & Cho, DS (2000) *Tiger Technology; The Creation of a Semiconductor Industry in East Asia*, Cambridge: Cambridge University Press.

Matsushita, M (1993) *International Trade and Competition Law in Japan*, Oxford: Oxford University Press.

Michie, J (ed.) (2003) *The Handbook of Globalisation*, Cheltenham: Edward Elgar.

Mikanagi, Y (1996) *Japan's Trade Policy: Action or Reaction?*, London: Routledge.

Mintzberg, H (1993) *Structure in Fives: Designing Effective Organisations*, New Jersey: Prentice Hall.

Montagnon, P (ed.) (1990) *European Competition Policy*, London: Pinter.

Mowery, DC (1995) 'Flat-panel displays', *Issues in Science and Technology*, Spring, p. 5.

Mowery, DC (1992) 'The US national innovation system: Origins and prospects for change', *Research Policy*, Vol. 21, No. 2, pp. 125–144.

Mowery, DC & Rosenberg, N (1989) *Technology and the Pursuit of Economic Growth*, Cambridge: Cambridge University Press.

Mowery, DC & Rosenberg, N (1989) 'New Developments in US Technology Policy: Implications for Competitiveness and International Trade Policy', *California Management Review*, Vol. 32, No. 1, pp. 107–124.

Murtha, TP, Lenway, SA & Hart, JA, (2001) *Managing New Industry Creation: Global Knowledge Formation and Entrepreneurship in High Technology*, Stanford, California: Stanford University Press.

Mytelka, L (2000) 'Localised Systems of Innovation in a Globalised World Economy', *Industry and Innovation*, 7, pp. 15–32.

Nature (1993) 'What road ahead for Korean science and technology', Vol. 364, 29 July, pp. 377–384.

Nembhard, JG (1996) *Capital Control, Financial Regulation, and Industrial Policy in South Korea and Brazil*, Westport, Connecticut: Praeger.

Neale, AD (1970) *Antitrust Laws of the United States of America: A Study of Competition Enforced by Law*, National Institute of Economic and Social Research, Cambridge: Cambridge University Press, 2nd edition.

Nelson, RR (ed.) (1993) *National Innovation Systems: A Comparative Analysis*, New York: Oxford University Press.

Nelson, RR (1990) 'US technological leadership: where did it come from and where did it go?', *Research Policy*, Vol. 19, No. 2, pp. 117–133.

Nelson, RR & Nelson, K (2002) 'Technology, institutions, and innovation systems', *Research Policy*, Vol. 31, No. 2, pp. 265–272.

New Scientist (1975) 'Calculator firms muscle into watchmaker's business', 22 May, p. 446.

Nihon Keizai (1997) 'Sony, Sharp, Philips to jointly develop next-generation large LCD', 30 July.

Nihon Keizai (1997) 'LCD makers increase capital spending', 24 March.

Nihon Keizai (1996) 'Philips to boost LCD sales in Japan to 120–130 billion yen ($1.07–1.16 billion) in fiscal 2000', 20 November.

Nihon Keizai (1996) 'Hosiden may establish TFT-LCD production joint venture with Philips', 11 October.

Nihon Keizai (1996) 'Three Korean firms to invest total of 230 billion yen ($2.07 billion) in LCD production expansion', 6 July.

Nikkan Kogyo (1997) 'Sharp, Sony, Philips develop 42-inch wide-screen plasma address LCD', 6 October.

Nikkan Kogyo (1997) 'Hosiden and Philips Display to increase LCD production capacity by 70 per cent in 1997', 29 May.

Nikkan Kogyo (1997) 'Sharp develops 20.1-inch TFT color LCD', 4 April.

Nikkan Kogyo (1996) 'Sharp to mass produce 15.0-inch TFT color LCDs next year', 30 September.

Nikkan Kogyo (1996) 'Sharp develops 40-inch TFT color LCD', 28 September.

Nikkan Kogyo (1996) 'Hosiden develops 23-inch TFT color LCD', 27 September.

Nikkan Kogyo (1996) 'Sharp develops 13.8-inch TFT-LCD with enhanced angle of field', 9 April.

Niosi, J, Saviotti, P, Bellon, B & Crow, M (1993) 'National Systems of Innovation: In Search of a Workable Concept', *Technology in Society*, Vol. 15, pp. 207–227.

Numagami, T (1996) 'Flexibility trap: a case analysis of US and Japanese technological choice in the digital watch industry', *Research Policy*, Vol. 25, No. 1, pp. 133–162.

O'Mara, W (1992) 'LCD Dividends', *Physics World*, Vol. 5, No. 6, June, p. 38.

Odagiri, H (1992) *Growth Through Competition, Competition Through Growth: Strategic Management and the Economy Through Growth*, Oxford: Clarendon Press.

OECD *Application of Competition Policy to High-Tech Markets*, GD/(97)44, Paris.

OECD *Economic Outlook* (various editions), Paris.

OECD *Globalisation of Industry: Overview and Sector Reports* (1992), Paris.

OECD *Competition Policy in OECD Countries 1992–93*, Paris: OECD (1995).

Ohmae, K (1990) *The Borderless World: Power and Strategy in the Global Market Place*, London: HarperCollins.

Okimoto, DI (1989) *Between MITI and the Market: Japanese Industrial Policy for High Technology*, Stanford, California: Stanford University Press.

Okimoto, DI, Sugano, T, Weinstein, FB, Flaherty, MT, Itami, H, Linvill, JG & Uenohara, M (1984) *Competitive Edge: The Semiconductor Industry in the US and Japan*, Stanford, California: Stanford University Press.

Park, WS (1994) 'The Recent Status of LCD Development in Korea', *The Electromechanical Society Proceedings*, Vols. 94–95, pp. 11–19.

Patel, P & Pavitt, K (1987) 'Is Western Europe losing the technological race'?, *Research Policy*, Vol. 16, Nos. 2–4, pp. 59–87.

Patel, P & Pavitt, K (1994) 'National Innovation Systems: Why They Are Important, And How They Might Be Compared', *Economics of Innovation and New Technology*, Vol. 3, pp. 77–95.

Pavitt, K & Patel, P (1996) 'What Makes High Technology Competition Different From Conventional Competition? The Central Importance of National Systems of Innovation' in G Koopman & HE Scharrer (eds) *The Economics of High-Technology Competition and Cooperation in Global Markets*, Baden-Baden: Nomos, pp. 143–171.

Perez, C & Soete, L (1988) 'Catching up in technology: entry barriers and windows of opportunity', in G Dosi *et al.* (eds) *Technical Change and Economic Theory*, London: Pinter, pp. 458–479.

Pecht, M, Bernstein, JB, Searle, D & Peckerar, M (1997) *The Korean Electronics Industry*, New York: CRC Press.

Pecht, M & Lee, CS (1997) 'Flat panel displays: What's Going On in East Asia Outside of Japan', CALCE Electronic Packaging Research Centre, University of Maryland.

Peters, SR (2000) *Competition and Technological Change in the Liquid Crystal Display Industry*, Unpublished Ph.D, Brunel University.

Pitkethly, R (2003) 'Analysing the Environment', in DO Faulkner & A Campbell (eds) *The Oxford Handbook of Strategy: Strategy Overview and Competitive Strategy*, Oxford: Oxford University Press, Vol. 1, pp. 225–260.

Porter, M (1996) 'What is Strategy', *Harvard Business Review*, November–December, pp. 61–78.

Porter, M (1990) *The Competitive Advantage of Nations*, London: Macmillan.

Porter, M (1980) *Competitive Strategy: Techniques for Analysing Industries and Competitors*, New York: The Free Press.

Prestowitz, C (1994) 'US needs initiative in flat-panel display', *Issues in Science and Technology*, Winter, p. 7.

Rosenbaum, DI & Lamort, F (1992) 'Entry, barriers, exit and sunk costs: An analysis', *Applied Economics*, Vol. 24, No. 3, pp. 297–304.

Rothwell, R & Zegveld, W (1985) *Reindustrialisation and Technology*, Harlow: Longmans.

Rosenberg, N (1994) *Exploring the Black Box: Technology, Economics and History*, Cambridge: Cambridge University Press.

Rosenberg, N (1982) *Inside the Black Box: Technology and Economics*, Cambridge: Cambridge University Press.

Ruttan, V (2001) 'Sources of Technical Change: Induced Innovation, Evolutionary Theory, and Path Dependence', in R Garud & P Karnoe (eds) *Path Dependence and Creation*, Mahwah, New Jersey: Lawrence Erlbaum Associates Publishers, pp. 91–123.

Rutton, VW (1997) 'Induced Innovation, Evolutionary Theory and Path Dependence: Sources of Technical Change', *Economic Journal*, Vol. 107, pp. 1520–1529.

Saviotti, PP (1997) 'Innovation Systems and Evolutionary Systems' in C Edquist (ed.) *Systems of Innovation: Technologies, Institutions and Organisations'*, London: Pinter, pp. 180–200.

Schmalensee, R (1981) 'Economies of Scale and Barriers to Entry', *Journal of Political Economy*, Vol. 89, pp. 1228–1237.

Schumpeter, J (1934) *The Theory of Economic Development: An Inquiry into Profits, Capital, Credit, Interest and the Business Cycle*, Cambridge, Massachusetts: Harvard University Press.

Schumpeter, J (1943) *Capitalism, Socialism and Democracy*, London: Allen & Unwin.

Scherer, FM (1994) 'Competing for Comparative Advantage Through Technological Innovation' in O Granstrand (ed.) *Economics of Technology*, London & Amsterdam: Elesevier Science, North-Holland.

Scherer, FM (1992) *International High-Technology Competition*, Cambridge, Massachusetts: Harvard University Press.

Scientific American (1991) 'Flat Horizons-US pursues research but little development of advanced screens', June, pp. 82–83.

SEC Family News (1997) 'Korea's TFT-LCD Business, Getting Brighter', Samsung Electronics, September, pp. 8–11.

Senge, P (1990) *The Fifth Discipline: The Art and Practice of the Learning Environment*, London: Century Business.

Sharp, M (1993) 'The Community and new technologies' in J Lodge (ed.) *The European Community and the Challenge of the Future*, London: Pinter, 2nd edition, pp. 200–223.

Sharp, M & Pavitt, K (1993) 'Technology Policy in the 1990s: Old Trends and New Realities', *Journal of Common Market Studies*, June, Vol. 31, No. 2, pp. 129–151.

Sharp, M & Pavitt, K (1991) 'The Single European Market and the Future of European technology policies' in C Freeman, M Sharp & W Walker (eds) *Technology and the Future of Europe: Global Competition and the Environment*, London: Pinter, 2nd edition, pp. 59–79.

Shin, JS (1996) *The Economics of the Latecomers: Catching Up, Technology Transfer, and Institutions in Germany, Japan and South Korea*, London: Routledge.

Sirower, M (1997) *The Synergy Trap: How Companies Lose the Acquisition Game*, New York: The Free Press.

The Economist (2005) Special Report, Samsung Electronics, 15 January.

The Economist (1995) 'Nuts, Bolts, Chips: Look at us, we can make it', 16 September.

The Economist (1994) 'Flat Screens: Crystal Diplomacy', 30 April, pp. 84–87.

The Economist (1993) 'Eenie, meenie, minie, mo...', 20 March, p. 106.

The Economist (1992) 'Gambling on flat screens', 15 August, pp. 71–72.

The Independent (1995) 'The commuter with a rolled-up screen', 13 November.

The New Republic (1993) 'How not to conduct trade policy: Flat panel flop', Vol. 29, No. 6, pp. 16–20.

The Nikkei Weekly (1993) 'Colour LCD makers stepping up output', 27 September.

Thompson, JL (2003) *Strategic Management*, London: Thomson Learning.

Tidd, J, Bessant, J & Pavitt, K (2005) *Managing Innovation: Integrating Technological, Market and Organisational Change*, Chichester: John Wiley & Sons, 3rd edition.

Tobias, M (1975) *An International Handbook of Liquid Crystal Displays*, London: Ovum.

Tsukada, T (1996) *Liquid Crystal Displays Addressed by Thin-Film Transistors*, Amsterdam BV: Gordon & Breach Publishers, Japanese Technology Reviews Vol. 29.

Tyson, L (1992) *Who's Bashing Whom: Trade Conflict in High-Technology Industries*, Washington DC: International Institute for Economics.

Valery, N (1975) 'Electronics in search of temps perdu', *New Scientist*, 30 October, pp. 284–85.

Valery, N (1975) 'Coming of age in the calculator business', Calculator Supplement, *New Scientist*, 13 November, pp. ii–iv.

Werner, K (1995) 'US Display Industry On The Edge', *IEEE Spectrum*, May, pp. 62–69.

Werner, K (1993) 'The flat panel's future', *IEEE Spectrum*, November, pp. 18–26.

Westphal, LE, Kim, L & Dahlman, CJ (1985) 'Reflections on the Republic of Korea's Acquisition of Technological Capability' in N Rosenberg & C Frischtak (eds) *International Technology Transfer: Concepts, Measures and Comparisons*, New York: Praeger, pp. 165–221.

Whitaker, DH & Kurosawa, Y (1998) 'Japan's crisis: evolution and implications', *Cambridge Journal of Economics*, Vol. 22, pp. 761–771.

Whittington, R (2001) *What is Strategy – And Does it Matter?*, London: International Thomson Learning Business Press, 2nd edition.

Wong, PK & Mathews, JA (1998) 'Competing in the Global Flat Panel Display Industry: Introduction', *Industry and Innovation*, Vol. 5, No. 1, pp. 1–10.

Woodward, OC & Long, T (1992) 'State of the Art', *Byte*, July, pp. 159–199.

Wilson, RW, Ashton, PK & Egan, TP (1980) *Innovation, Competition and Government Policy in the Semiconductor Industry*, Massachusetts: Lexington Books.

Wu, I Wei (1994) 'High-Definition displays and technology trends in TFT-LCDs', *Journal of the Society for Information Display*, Vol. 2, No. 1, pp. 1–14.

Index

259